Solar Cookers and Sustainable Cooking Solutions for the Post-Energy Crisis Era

Khalil Kassmi
https://orcid.org/0000-0002-4808-873X
Mohamed First University, Oujda, Morocco

Olivier Deblecker
https://orcid.org/0000-0003-2156-9450
Université of Mons, Belgium

IGI Global Scientific Publishing
Publishing Tomorrow's Research Today

Vice President of Editorial Melissa Wagner
Director of Acquisitions Mikaela Felty
Director of Book Development Jocelynn Hessler
Production Manager Mike Brehm
Cover Design Jose Rosado

Published in the United States of America by
IGI Global Scientific Publishing
701 East Chocolate Avenue
Hershey, PA, 17033, USA
Tel: 717-533-8845 | Fax: 717-533-7115
Website: https://www.igi-global.com E-mail: cust@igi-global.com

Library of Congress Cataloging-in-Publication Data

Names: Kassmi, Khalil editor | Deblecker, Olivier, 1971- editor
Title: Solar cookers and sustainable cooking solutions for the post-energy crisis era / edited by Khalil Kassmi, Olivier Deblecker.
Description: Hershey, PA : IGI Global Scientific Publishing, [2026] | Includes bibliographical references and index. | Summary: "This book aims to provide an in-depth analysis and a framework for reflecting on sustainable cooking solutions, with a particular focus on solar cookers"-- Provided by publisher.
Identifiers: LCCN 2025019286 (print) | LCCN 2025019287 (ebook) | ISBN 9798337325354 hardcover | ISBN 9798337325361 paperback | ISBN 9798337325378 ebook
Subjects: LCSH: Solar ovens | Solar cooking
Classification: LCC TS425 .S65 2025 (print) | LCC TS425 (ebook) | DDC 621.47--dc23/eng/20250729
LC record available at https://lccn.loc.gov/2025019286
LC ebook record available at https://lccn.loc.gov/2025019287

British Cataloguing in Publication Data
A Cataloguing in Publication record for this book is available from the British Library.

Table of Contents

Detailed Table of Contents

A previous Journal publication from the authors addressed the theoretical background of a minimum Photovoltaic Solar Cooker (mPVSC) that operates without batteries or electronics. It is suitable for low-income communities and remote or isolated locations. It operates as a stand-alone cookware, allowing cooking indoors in a reasonable timeframe, thus improving the quality of life of the cook and family. It is robust, easily deployed, fail-safe, and offers renewable electrical energy for other basic needs PV panel power is maximized by using a set of manually operated switches and an electric load formed by a plurality of Positive Temperature Coefficient (PTC) heaters, which are now commercially available at low cost. A prototype showcasing this design has been tested under outdoor conditions. Its performance was evaluated by applying heat to a load. Additionally, heat losses were quantified during the cooling periods.

Solar energy is increasingly vital in global strategies to reduce greenhouse gas emissions. Solar cooking is a key application, especially in rural areas where it lowers reliance on wood, improving health and reducing deforestation. This study focuses on designing, simulating, building, and testing a solar box cooker. Thermal performance was simulated using TRNSYS, followed by outdoor tests in Agadir, Morocco, under varying conditions—different water loads, with/without reflectors, and with/without thermal load. A solar tracking system was implemented to keep the cooker aligned with the sun, improving efficiency. The prototype achieved a strong first figure of merit F1 (0.124 $m^2{}^\circ C/W$), indicating high optical efficiency, and a second figure F2 (0.436), showing effective heat transfer. The cooker reached energy and exergy efficiencies of 19.36% and 10.12%, respectively. Overall, the tracking system significantly enhanced the performance of the solar box cooker.

This chapter presents the digitization and supervision of a photovoltaic solar cooker using a Raspberry Pi Pico W microcontroller. The system collects meteorological

reactions for long-term heat retention with minimal losses. A comparative analysis highlights their thermal properties, cost effective, and practicality for solar cooking. Effective TES integration requires high energy density, thermal stability, durability, and ease of use, and this chapter discusses how TES systems can be tailored to various solar cooker designs, from small household units to large-scale community systems. The discussion also covers key aspects such as thermal insulation, heat retention, and heat transfer mechanisms to maximize efficiency and minimize heat loss, while safety and user-friendliness are emphasized for adoption in resource-limited settings. TES significantly enhances solar cooking by extending cooking hours, reducing reliance on conventional fuels, and improving user convenience. However, challenges such as heat loss, material degradation, and inefficiencies, along with economic and accessibility barriers, must be addressed. The chapter underscores the need for continued research and innovation to improve TES-integrated solar cookers, making them more affordable, efficient, and accessible for widespread adoption.

The growing use of fossil fuels has led to a significant rise in greenhouse gas emissions, directly contributing to global warming and environmental imbalances. The burning of coal, oil, and natural gas releases large amounts of carbon dioxide (CO_2), methane (CH_4), and nitrogen oxides (NO_X), intensifying the greenhouse effect and accelerating climate change. As a result, transitioning to renewable energy sources is crucial to mitigate global warming, protect public health, and preserve the environment. Solar cookers are an effective technology that captures and converts solar radiation into usable heat for various applications. This article presents a thermal study on three solar cooker designs: rectangular, trapezoidal, and triangular. The goal is to determine the temperatures of the absorber surface and indoor air. The study compares the effects of single and double glazing on thermal performance and considers the impact of adding or removing a thermal load inside the cooker. Additionally, simulations with a primary reflector are also analyzed to assess its effect on performance.

Energy conservation has been a primary focus for scientists. A significant issue of the present is the fast depletion of earth's resources. It is important to diminish reliance on non-renewable resources. Efforts in solar cooking have consistently shown their advantages and efficacy. Solar cooking is a technology that derives energy from photovoltaic solar panels driven by sunshine. We possess a technique that stores solar energy in a battery pack for use during periods without sunlight. Electrical energy is generated by photovoltaic panels (PV) and stored in a battery. Consequently, the cookers are energized by boost converter. The integration of photovoltaic panels and batteries will provide the necessary electricity during daylight hours. A comprehensive primary grid backup is also included into the system. This article investigates electric cooking system aimed at advancing renewable and sustainable cooking technology with the application of power electronic converters

The main objective of this work is to provide an overview of used solar cookers and to describe the essential parameters for testing and evaluating their thermal performance. Several categories of solar cookers are found in the literature servey. The price of solar cookers has recently been reduced by around 56%, from 530 Dhs to 235 Dhs. The manufacturing cost remains around 200 Dhs. They are classified into four main families: box cookers, concentrator cookers, ovens, and indirect solar cookers. Each family is divided into three types of solar cookers namely: Photovoltaic, Thermal and Hybrid solar cookers. Photovoltaic solar cooker remains the cooker of the future.

Preface

In many households around the world, cooking still relies on traditional fuels such as wood, charcoal, or gas, as well as electricity generated from fossil sources. These energy sources now pose serious challenges in terms of cost, availability, and sustainability. In some regions, particularly in Africa, people continue to depend on forest wood for heating and cooking. Such practices not only deplete natural resources for future generations but also contribute to environmental degradation, leading to phenomena such as droughts and floods.

To address these challenges, it has become essential to turn to renewable energy sources—abundant, free, and environmentally friendly. All energy applications, including cooking, must now move toward these sustainable alternatives. In this context, the development of solar cooking emerges as a priority solution, encompassing both traditional thermal cookers (box-type, concentrating systems, etc.) and electric cookers powered by photovoltaic energy.

Solar cooking thus represents a key field of innovation in the global context of energy transition and sustainable development. In light of environmental concerns and the persistent dependence on fossil fuels, solar-based cooking technologies offer practical, economical, and ecological solutions suitable for both developed and developing regions.

This book explores various technological approaches to improve the efficiency and flexibility of solar cookers by combining traditional thermal methods with modern photovoltaic systems. The contributions presented cover a wide range of research and applications—from simple, minimalist solar cookers to hybrid devices that integrate thermal storage and energy optimization through artificial intelligence and the Internet of Things (IoT).

The volume also emphasizes the design, simulation, experimental evaluation, and performance optimization of solar cookers in different climatic and geographical contexts, enabling readers to understand the technical challenges and innovative solutions developed by researchers worldwide.

As such, this compilation serves as a comprehensive reference for engineers, researchers, students, and doctoral candidates, as well as for policymakers interested in the development of sustainable solar technologies and the promotion of environmentally friendly energy solutions. It illustrates how solar cooking can help reduce energy poverty, enhance food security, and support the achievement of sustainable development goals. Furthermore, this book stands as a valuable educational and scientific resource, offering students and researchers a complete guide to the concepts, experimental methods, and innovations in the field of solar cooking.

Today, the world faces major energy and environmental challenges. A large portion of the global population continues to rely on fossil fuels and firewood for cooking, leading to increased deforestation, air pollution, and greenhouse gas emissions, while exacerbating energy poverty in vulnerable regions. Even in urbanized areas, electricity is often generated from non-renewable sources, contributing to fossil fuel dependence and the volatility of energy costs.

In this context, solar cooking represents a relevant and sustainable solution. It fully aligns with global energy transition policies and the United Nations Sustainable Development Goals (SDGs), particularly those related to clean energy, food security, and climate change mitigation. Solar cookers offer multiple advantages: reducing fossil fuel consumption, decreasing pollutant emissions, lowering household energy expenses, and strengthening local energy security.

- Recent innovations go beyond traditional solar cookers and include:
    concentrating or box-type thermal cookers adapted for both domestic and community needs;
- photovoltaic-based electric cookers integrating energy storage and intelligent control, allowing continuous use even during non-sunny periods;
- smart solar cooking systems whose heating and regulation adapt to users' modern cooking habits, ensuring optimal acceptability of this new technology;
- optimization through artificial intelligence (AI) and the Internet of Things (IoT) to maximize energy efficiency and adapt performance to local climatic conditions.

Thus, solar cooking is not limited to regions with high energy vulnerability but also represents a field of global technological and scientific innovation. Researchers, engineers, and policymakers increasingly recognize the importance of developing efficient, affordable, and locally adapted systems within the broader framework of sustainability and energy transition.

In summary, solar cooking has emerged as a credible, efficient, and environmentally friendly alternative that addresses crucial social, economic, and ecological challenges. It also serves as a valuable educational and scientific resource for students, doctoral candidates, and professionals dedicated to advancing sustainable energy solutions.

This book is intended for a wide range of readers interested in solar cooking and related technologies. It primarily serves as a solid foundation for understanding this emerging mode of energy use, enabling readers to become familiar with the fundamental principles, technological innovations, and design approaches of solar cookers.

It also provides an overview of the state of the art, current topics, and key research directions in the field, thus offering students, doctoral candidates, and researchers a valuable educational and scientific tool to deepen their knowledge and guide future studies.

The book stands out for its interdisciplinary approach, combining solar energy, engineering, materials science, electronics, artificial intelligence, the Internet of Things (IoT), and environmental sciences. It includes experimental studies, simulations, field tests, and comparative assessments, allowing readers to understand the technical challenges and innovative solutions developed around the world.

For engineers, industrial stakeholders, and technology developers, this book presents opportunities for the industrialization and implementation of eco-friendly solar cookers adapted to both rural and urban households. It emphasizes the design of user-acceptable devices, including intelligent regulated cookers aligned with modern cooking habits, thus ensuring the adoption and efficiency of these new technologies.

Finally, this book serves as a comprehensive reference for policymakers, NGOs, and development practitioners, enabling them to promote sustainable, accessible, and locally adapted energy solutions while contributing to energy transition, food security, and environmental protection goals. With its practical and theoretical dimensions, it represents a valuable resource for education, applied research, and the development of pilot projects at both local and international levels.

To better understand the major innovations behind this new generation of solar cookers—whether thermal or electric, intelligently regulated to adapt to modern user habits, and integrating energy storage and optimization solutions—the chapters presented in this book offer a comprehensive and diversified exploration. Each contribution presents advanced research, innovative technologies, and practical applications that demonstrate how solar cooking can sustainably transform domestic practices, reduce environmental impact, and contribute to the global energy transition.

In order to provide a complete and well-structured overview of this rapidly expanding field, the following chapters are included:

CHAPTER OVERVIEWS

Chapter 1 – Minimal Photovoltaic Solar Cooker With No Battery or Electronics. Testing and Assessment: A tool to fight energy poverty and vulnerability

This chapter presents a minimal photovoltaic solar cooker (mPVSC) operating without batteries or electronics, designed for low-income communities and remote areas. This simple and robust system allows indoor cooking while providing electricity for other household needs. The tested prototype demonstrates that it is possible to eliminate electronic components while maintaining reliable performance, thereby reducing costs and fire risks.

Experiments show that the cooker can bring 1 kg of water to a boil within a reasonable time, even with a non-insulated pot, and that the use of an insulating sleeve improves heat retention and cooking duration. The concept can be extended to community installations for schools, hospitals, or community centers.

This chapter provides engineers and policymakers with a practical tool to combat energy poverty and vulnerability while offering a pathway toward affordable and sustainable solar systems. It also inspires the design of future minimalist photovoltaic and electric cookers.

Chapter 2 – Performance Evaluation of an Inclined Solar Box Cooker under Different Climatic Conditions of Agadir City: Design, Simulation and Experimental Testing

This chapter presents the design, simulation, and experimental testing of an inclined solar box cooker optimized for the climatic conditions of Agadir, Morocco. The prototype was simulated using TRNSYS and then tested outdoors with varying water loads, the use of reflectors, and a solar tracking system to maintain the cooker's orientation toward the sun.

The results show excellent optical efficiency (F1 = 0.124 m^2°C/W) and good heat transfer efficiency (F2 = 0.436), with energy and exergy efficiencies of 19.36 % and 10.12 %, respectively. The optimal tilt angle approximately corresponds to the local latitude, and the combination of solar tracking with reflectors significantly improves performance.

This chapter highlights the importance of geometric design and orientation technologies to maximize solar cooker efficiency. The methods and results presented provide a valuable guide for engineers and researchers seeking to develop solar cooking solutions adapted to local conditions and sustainable energy needs.

Chapter 3 – Innovation and Digitization of Solar Cookers Towards Sustainable Energy Management

This chapter presents the intelligent supervision and regulation of a photovoltaic solar cooker using a Raspberry Pi Pico W microcontroller. The system collects real-time weather and electrical parameters (voltage, current, power, efficiency, temperatures) and automatically adjusts the cooker's operation. Data, transmitted via the MQTT protocol, are displayed on a Tkinter interface and stored in a MySQL database, ensuring complete traceability.

- The results demonstrate that this architecture enables:
  real-time visualization and timestamped data storage for individualized monitoring,
- deferred analysis through an interactive web application,
- reliable adaptation of the solar cooker to different contexts.

Laboratory and field tests confirm that the system combines autonomy, intelligence, and energy performance. This approach provides a practical solution for the energy transition in rural areas, enhancing user comfort and safety while reducing environmental impact.

Chapter 4 – Integration of Thermal Energy Storage Materials for Off-Sunshine Cooking

This chapter explores the integration of thermal energy storage (TES) into solar cookers, a key solution for overcoming direct dependence on sunlight. TES allows excess thermal energy captured during sunny hours to be stored and used for cooking during periods without sun, such as evenings, nights, or cloudy days.

- TES systems are classified into three main types:
- Sensible heat storage: materials such as water, oil, or rocks that store and release heat through temperature variation.
- Latent heat storage: phase change materials (PCMs) such as paraffin wax or hydrated salts, which store and release energy during phase transitions.
- Thermochemical storage: reversible chemical reactions providing long-term storage with minimal losses.

The chapter also details the optimization of TES according to the type of cooker, emphasizing thermal insulation, heat retention, heat transfer, safety, and ease of use.

TES integration significantly enhances the practicality, efficiency, and reliability of solar cookers, reducing dependence on fossil fuels and improving user comfort.

The authors nevertheless highlight challenges related to heat losses, material degradation, and high initial costs. Ongoing research and innovation, combined with supportive policies and user awareness, are essential to promote widespread adoption of TES-equipped solar cookers, from domestic households to community and industrial applications.

Chapter 5 – Study of the Thermal Performance of Different Solar Cooker Geometries with Single and Double Glass: CFD Simulation of Different Solar Cooker Designs with Simple and Double Glazing

This chapter analyzes the thermal performance of different solar cooker geometries (rectangular, trapezoidal, triangular) using CFD simulations. The study compares the effect of single and double glazing, internal heat load, and the addition of primary reflectors or angled mirrors on the concentration of solar radiation and the rise in temperature at the surface of the absorber.

The results show that:

1. The geometry of the cooker directly influences the solar concentration and the stagnation temperature of the absorber.
2. Double glazing significantly improves thermal efficiency, reduces cooking time and optimizes load capacity by limiting heat loss thanks to the insulating air layer.
3. The addition of reflectors and the positioning of the internal mirrors further increase efficiency, reducing heat losses and increasing the F1 and F2 figures of merit.

This work highlights the importance of geometric design, glazing and reflective systems to maximize heat transfer to the baking chamber. It provides essential indicators for comparing and optimising different types of solar cookers and provides a practical basis for designers and researchers wishing to improve the energy efficiency of solar home and community systems.

Chapter 6 – Solar Clean Cooking System with the Aid of Power Converters: Towards Sustainable Cooking Revolution

This chapter presents an innovative clean solar cooking system integrating photovoltaic panels, batteries, and power converters to supply electricity to cookers even in the absence of sunlight. This approach combines sustainable cooking with operational flexibility, addressing the needs of both urban and rural households. The system emphasizes simplicity, reliability, and energy efficiency while reducing the ecological footprint compared to conventional methods. Future research envisions the integration of artificial intelligence and the Internet of Things (IoT) to optimize energy management and cooker design, as well as the development of thermal and hydrogen storage solutions. This chapter illustrates the potential of photovoltaic and electronic technologies to transform domestic cooking and contribute to a sustainable revolution in the culinary sector.

Chapter 7 – Synthesis of Solar Cookers and Evaluation of Their Thermal Performances.

This chapter provides a comprehensive overview of solar cookers used around the world and describes the essential parameters for evaluating their thermal performance. Solar cookers are classified into four main categories: box cookers, concentrating cookers, solar ovens, and indirect solar cookers, with each category further subdivided into thermal, photovoltaic, and hybrid cookers.

The focus is on photovoltaic cookers, considered the cooking technology of the future. These devices allow higher cooking temperatures by adjusting PV generator parameters and offer the possibility of storing energy in solar batteries to cook even in the absence of sunlight. The chapter also presents the evolution of costs, highlighting a recent reduction of approximately 56% in cooker prices, making this technology more accessible.

Finally, the study includes a literature review and a thermal performance analysis, emphasizing the ecological and economic benefits of solar cooking, as well as its potential to reduce dependence on fossil fuels and limit CO_2 emissions. Technical improvements, such as the use of movable mirrors in concentrating cookers, are also discussed to optimize thermal efficiency and overall performance.

Chapter 8 – Health, Safety, and Sustainability Benefits of Solar Cooking

This chapter examines solar cooking as an innovative and transformative solution for food preparation, offering significant health, safety, and environmental benefits. By eliminating indoor air pollution and reliance on hazardous fuels, solar cookers reduce respiratory and cardiovascular risks while improving the nutritional retention of food.

From an environmental perspective, solar cooking contributes to sustainability by limiting deforestation, CO_2 emissions, and energy consumption. Advances in solar cooker design enhance efficiency, safety, and accessibility, making them suitable for households worldwide. The chapter also highlights the importance of addressing challenges related to weather conditions, cost, and cultural acceptance through innovation, training, and policy support.

Finally, the chapter paves the way for future research aimed at optimizing systems for diverse climates, improving thermal storage for night-time or low-sunlight cooking, and evaluating long-term health impacts. The integration of solar cooking with hybrid systems and supportive policy frameworks is identified as a promising pathway for global and sustainable deployment.

This book represents a significant milestone in the advancement of solar cooking, combining an in-depth literature review of all types of cookers with experimental and simulation results derived from practical research. This dual approach, both fundamental and applied, transforms knowledge into practical and innovative solutions.

Through the integration of advanced technologies—phase change materials, photovoltaic cookers, IoT-supervised systems, and PSO-based optimization—the book provides concrete tools to enhance thermal efficiency, extend energy autonomy, and maximize the performance of solar cookers.

Readers will find a solid foundation for research and development, as well as practical recommendations for the design and adoption of sustainable solutions. Engineers, policymakers, rural communities, and researchers can implement solar cookers that are efficient, safe, and accessible, contributing to reduced carbon emissions, the preservation of natural resources, and improved living conditions.

Ultimately, this book goes beyond compiling knowledge: it paves the way for innovation, applied research, and the global dissemination of sustainable solar cooking, empowering each reader to become a potential contributor to the energy and environmental transition.

Acknowledgements

As editors, we wish to express our deep gratitude to all the authors who contributed to this book. Their commitment, scientific rigor, and the quality of their work have made it possible to produce chapters that are current and relevant, covering a wide range of fundamental and applied research on solar cookers and sustainable cooking.

We also extend our thanks to everyone who participated in the exchanges, discussions, and collaborations throughout this project. These interactions were essential for structuring the chapters, sharing knowledge, and providing a comprehensive overview of the field. The simulation results, experimental studies, and analyses presented in the various chapters provide a solid foundation for future work and innovation in this domain.

This book thus reflects a collective effort that combines theoretical knowledge, applied research, and practical perspectives, offering readers a valuable resource to understand, develop, and improve solar cooking technologies. We hope this contribution will serve as a reference and stimulate new research, collaborations, and applications in the field.

Chapter 1
Minimal Photovoltaic Solar Cooker With No Battery or Electronics:
Testing and Assessment

Antonio Lecuona-Neumann
https://orcid.org/0000-0002-5763-4982
Instituto Universitario de Desarrollo Tecnológico y Promoción de la Innovación Pedro J. De Lastanosa, Spain

Jean Boubour
https://orcid.org/0009-0009-3141-0225
Independent Researcher, France

Maria Del Carmen Rodriguez-Hidalgo
https://orcid.org/0000-0001-5379-0577
E.T.S.I. Navales, Spain

Jose Ignacio Nogueira-Goriba
https://orcid.org/0000-0003-1866-6110
INAIA, Spain

ABSTRACT

A previous Journal publication from the authors addressed the theoretical background of a minimum Photovoltaic Solar Cooker (mPVSC) that operates without batteries or electronics. It is suitable for low-income communities and remote or isolated locations. It operates as a stand-alone cookware, allowing cooking indoors in a reasonable timeframe, thus improving the quality of life of the cook and family. It is robust, easily deployed, fail-safe, and offers renewable electrical energy for

DOI: 10.4018/979-8-3373-2535-4.ch001

other basic needs PV panel power is maximized by using a set of manually operated switches and an electric load formed by a plurality of Positive Temperature Coefficient (PTC) heaters, which are now commercially available at low cost. A prototype showcasing this design has been tested under outdoor conditions. Its performance was evaluated by applying heat to a load. Additionally, heat losses were quantified during the cooling periods.

1. INTRODUCTION

According to the World Health Organization (WHO) in 2022, globally, around 2 billion individuals depend on firewood, dung, agricultural residues, and similar solid fuels for their needs. This traditional practice is deeply embedded in various cultures. Fire offers warmth and nourishment in households throughout developing regions and is characterized by a low thermal efficiency. Using firewood for cooking brings considerable drawbacks, such as health hazards from indoor air pollution, deforestation, and the time lost and the trauma associated with carrying heavy loads of collected wood across long distances (Geng & Bai, 2023), (Bruce, Perez-Padilla, & Albalak, 2000). For communities seeking sustainable alternatives, transitioning to cleaner cooking technologies is crucial. This shift aims to enhance health outcomes, improve quality of life, and safeguard our planet for future generations.

The transition to improved solid fuel stoves and fossil fuels marks a significant advancement, primarily due to their lower emissions and enhanced efficiency (Bailis, Drigo, A., & Masera, 2015). However, this progress appears insufficient (Das, 2025). The introduction of thermal solar stoves completely addresses the issue of air pollution, either local or global (Sarangi, Sarangi, Sahoo, Nayak, & Mallik, 2024), (Arunachala & Kundapur, 2020), (Kundapur, 2018). A more sophisticated solution involves using electric cookers. For this last option, in situations where extending the grid to the consumption point is not economically viable, alternative solutions may be necessary, particularly for subsistence use or rural deployment. Locally produced electricity by Photo-Voltaic panels (PV) (Yayhyaoui, 2018) allows reaching small and remote communities. This approach does not only foster energy independence but also serves as the sole alternative for remote and isolated communities. Solar energy is widely available across very wide regions of the globe, particularly in subtropical areas where the need for modern cooking appliances is most acute. This is the case for the Sagel region and parts of Asia, among others. (Aemro, Moura, & de Almeida, 2021), (De, Shawhatsu, De, & Ajaero, 2013) describe solar cooking as a healthy and environmentally caring cooking method. Additionally, PV cookers enable indoor cooking, preserving traditional culinary practices while allowing the home keeper to manage other responsibilities. Despite the large availability of solar energy, its intermittency precludes relying only on solar cooking.

In sufficiently large communities, smart grids facilitate the sharing of large PV fields through AC electric distribution, eventually connected to the grid as a backup, allowing electric cooking (Khalifa & Al, 2025). However, for smaller communities and impoverished peri-urban areas, single family PV fields emerge as a more viable solution (Sillé & van Duijsen, 2023). Attempts have been performed to allow indoor solar cooking, but none of them exhibit the flexibility of PV (Harmim, Merzouk, & Boukar, 2013), (Kaushik & Gupta, 2008), among others. Moreover, with PV oriented to cooking additional use, such as electrification, is made possible. The nominal power needed for full PV family cooking, 200 to 1.000 W, is large enough to allow charging low-power devices, such as flashlamps, radios, and phones (Rahul Kashyap, Pramanik, & Ravikrishna, 2023).

A PV individual or family system typically includes PV panels with a power maximizer (Maximum Power Point Tracker MPPT), a battery charger for energy storage, and voltage regulation. Optionally, it includes an inverter for converting Direct Current (DC) to Alternating Current (AC) to facilitate the use of conventional home appliances (Atmane, Kassmi, Deblecker, & Bachiri, 2021). However, a simpler and less expensive option is possible. For small electrical circuits, AC is often unnecessary and impractical due to operational constraints, added losses, and cost. Furthermore, batteries can be expensive, have a limited lifespan, and can pose environmental hazards if not disposed of properly at the end of their operational life (Lecuona-Neumann, Nogueira, & Legrand, 2018). Nowadays, the low cost of PV panels enables proposals for indoor solar cooking without electronics and batteries, such as the experimental platform in (Vipin, Singh, & Yadav, 2025), affordable by economically vulnerable communities with the appropriate financial help.

In cooking, the essential skill lies in effectively harnessing Thermal Energy (TE) at temperatures exceeding about 70 °C. Temperature breaks down starch molecules into more digestible fragments. It "denatures" protein molecules so that their amino acid chains are broken, so that digestive enzymes can attack them more easily. It takes time to reach elevated temperatures in the interior of solid food pieces, which is necessary to guarantee the elimination of pathogens. For that, maintaining elevated temperatures for some time is necessary. Extra heat must be applied to compensate the losses to ambient from the cooking utensil. To avoid further consumption, the heat can also be retained (Thermal Energy Storage or TES). There are two possibilities.

In the first possibility, the food itself acts as TES to prolong cooking. (Lecuona, Nogueira, Ventas, Rodríguez-Hidalgo, & Legrand, 2003). After the initial cooking phase, if the heated food is adequately thermally insulated, it can continue to cook at progressively lower temperatures until it reaches the End of Cooking (EoC) point, which is around 60°C. This method is known as slow cooking or retained heat cooking. This technique also enables the food to stay warm and ready for consumption for an extended period, from when the sun is stronger at lunchtime through dinner and even overnight. This TES replaces batteries for cooking.

The other possibility is to use an extra mass as TES, either as sensible heat storage (e. g. water, rocks, …) or as latent heat storage in what is called a Phase Change Material (PCM), such as paraffins, sugar alcohols, hydrated salts, among others. Some examples of sensible heat storage are (Lugolole, Mawire, Lentswe, Okello, & Nyeinga, 2018) using oil, (Vipin, Singh, & Yadav, 2025) using cast iron so that higher temperatures are allowed, although other substrates have been applied, such as metals, pebbles, sand, or a mixture of them. PCMs increase the heat storage density at the expense of a higher cost and a shorter lifetime e. g. (Mawire & Abedigamba, 2025), (Lecuona, Nogueira, Ventas, Rodríguez-Hidalgo, & Legrand, 2003).

Further investigation is necessary to tackle practical challenges, including acceptability, scalability, cost-effectiveness, and long-term performance. Future research should prioritize overcoming these obstacles, particularly in the integration of PCMs with renewable energy systems such as solar cookers. This focus will significantly contribute to the advancement of energy-efficient and sustainable technologies that address primary needs in energy-vulnerable communities.

The Maximum Power Point Tracker (MPPT) electronics plays a vital role in the operation of PV panels. These panels require a specific operating voltage that fluctuates with changes in solar irradiance and temperature.

The MPPT function is the key responsibility of a complex digital electronic circuit. For remote, isolated, and economically vulnerable communities, the idea of eliminating electronic circuits is appealing, as it reduces complexity, drops costs, increased robustness, and minimizes the risk of breakdown of an essential commodity.

In the context of family PV cookers, there have been efforts to integrate non-linear load resistances, with an intrinsic function to approach MPPT, as noted by Ossei & Al (2024), in this case including a PCM load for TES.

Lecuona-Neumann et al. (2024) describe a minimal Photovoltaic Solar Cooker (mPVSC) with no electronics and no batteries. In this system, the user manually approaches the MPP by visually monitoring a low-cost DC power meter to determine the optimal number of the available heating resistances that electrically load the PV panel. To compensate for power loss because of no automatic MPPT, the aperture area of the PV panel can be increased if needed. Remarkably, the cost of these panels has significantly decreased, now ranging from € 0.1 to € 0.2 per watt-peak wholesale, making them an accessible solution, even counting with a low efficiency.

Additionally, Lecuona-Neumann et al. (2024) offer a theoretical study on a mPVSC that goes a step further. A set of Positive Temperature Coefficient (PTC) low-cost commercial ceramic resistances are used to avoid any risk of fire as they automatically limit their temperature to about 250 °C. The resulting voltage of a domestic PV panel of about 30 to 40 V DC precludes any risk to human health because of electrical shock. To keep this safety measure, additional identical PV panels must be connected in parallel.

Our research indicates that no experimental studies have been published featuring a design like the one presented here.

1.1. Aim and Scope

The primary objective of this study is to collect experimental data from a minimal Photo-Voltaic (PV)-driven Solar Cooker (mPVSC) prototype that operates without electronics or batteries in a representative configuration under sunlight. This serves as a foundation for future development. This unique combination could significantly transform energy usage and consumption at remote and isolated communities.

This research introduces a novel approach for acquiring crucial thermal data via a heating process, subsequently followed by a cooling process, under various operating conditions.

The relevance of this research stems from its potential for practical application, utilizing a demonstrative and innovative low-cost platform alongside straightforward experimentation.

This research contributes significantly to the growing body of knowledge on energy-efficient, sustainable systems. The results not only advance the theoretical understanding of PTC applications as ohmic resistances in cooking but also offer practical insights into their real-world benefits. This study provides a valuable reference for researchers and industry practitioners aiming to develop energy-saving technologies and contribute to a greener, low-carbon future.

This paper presents experimental results that provide a proof for a previous theoretical study; thus, it serves as a practical demonstration. They also offer insight, not only into the performance of the mPVSC implemented, but also offer critical comparisons with traditional cooking methods and new tools for analysis. Additionally, the study highlights the benefits of employing a type of replaceable thermal insulation that minimizes heat loss to the atmosphere during cooking compared to operating without this insulation.

Water-boiling experiments serve as a standard method for characterizing solar cookers, as in (American Society of Agricultural and Biological Engineers, 2013), (Bureau of Indian Standards (BIS), 1992), (Bureau of Indian Standards (BIS), 2000). (Lahkar & Samdarshi, 2010) offers a detailed review of standardized testing a kind of thermal cookers, the box cooker, which is partially applicable to our purposes. It is not entirely applicable because the PTCs have the so-called Curie temperature, T_C. At this temperature, PTCs progressively limit the input power as a solid-state thermostat by reversibly increasing, exponentially, its internal electrical resistance

R, because of a phase change in the PTC material (Kinyua, Wambugu, & da Silva, 2025). In this paper, in addition to the boiling test, a cooling test is proposed and its results discussed under the light of the lumped parameter model, taking advantage of multi temperature measurements. Moreover, the data gathered allows us to propose improvements for practical designs of the mPVSC.

Utilizing this mPVSC to heat water to boil, or oil to avoid evaporation heat loss, and allowing higher temperatures that are representations of food cooking completes the testing campaign. Subsequently, leaving the whole utensil cool down naturally, the overall heat conductance and the heat transfer coefficient U from the cooking pot to the atmosphere can be calculated. The data gathered characterizes the utensil plus pot efficiency and the time available for residual cooking. This information is valuable, not only for the design of photovoltaic cookers, but also for traditional cookers (Cardoso, et al., 2023; Mawire & Abedigamba, 2025). Furthermore, it evaluates the effects of integrating thermal insulation into cooking pots, akin to the practices currently employed in commercial Electric Pressure Cookers (EPC) (Cardoso et al., 2023; Mawire & Abedigamba, 2025).

The layout of the mPVSC under discussion is quite simple: it consists of a flat hot plate heated by dissipating the PV panel electricity through PTCs, here named the cooking utensil. It is complemented with a removable cooking pot positioned on top. The PTCs are of the flat type, Figure 1. The slab of ceramic material offers a frontal area specific resistance [$\Omega\ m^{-2}$]. For reducing resistance, a larger area can be devised for as the material acts in parallel, either with larger PTCs elements or with a larger number of them. They PTCs withstand elevated temperatures and support large voltages prior to breakdown. A removable thermal insulating jacket enhances the mPVSC.

The dimensions of the utensil were previously established for a smaller PV panel. During the experimental campaign, although successful performances were attained, several design flaws were identified, leading to proposed amendments. This, however, highlights the considerable flexibility of the mPVSC concept.

Section 2 of this paper outlines the model used to postprocess the experimental results from the testing platform, PV panel, utensil, pot, insulation and instrumentation, which is also described in this section. Section 2 also introduces the parameters of interest obtained from the experiments, the theory under them and the limitations encountered for their application to the mPVSC. Different performance standards are discussed and their application evaluated. Section 3 presents the results obtained after the testing campaign, accompanied by commentary and suggestions for improvements, while Section 4 provides the conclusions.

2. MATERIALS AND METHODS

2.1 Mathematical Modeling

The theory needed for processing the experimental data is exposed in this section, starting with the cooling experiment after heating and eventually boiling water. Afterwards, the algorithm resulting from mathematically modeling the preceding heating period is exposed, indicating its limitations and significance.

2.1.1. Cooling Process

For the cooling process, after heating the cooking utensil and pot, a single body lumped temperature T evolution can be assumed. It can be checked that the order of magnitude for the time to reach a fairly homogeneous temperature distribution across the body, according to the Laplacian equation of temperature for thermal conduction inside an homogeneous body, is $t_e \sim L^{*2} \alpha^{-1} \sim 10^{-2}\ \mathrm{m}^2 \times \left(10^{-4} \frac{\mathrm{m}^2}{\mathrm{s}}\right)^{-1} \sim 100$ s. Here L^* is a characteristic length of the cooking utensil plus pot and $\alpha = k\left(\rho c_p\right)^{-1}$ is its thermal diffusivity, which here is taken for a metal. t_e is short in comparison with the characteristic times for heating t_h and cooling $t_{coo} \sim 10^3$ to 10^4 s, defined in Eq. (7). After this time differences, internal isothermicity can be acceptable if the whole body Biot number is much smaller than unity, $Bi = UL^* k^{-1} \sim (1 \text{ to } 10) \times 10^{-1} \times 10^{-2} \ll 1$, where the conductivity k of a metal has been used (Incropera, DeWitt, Bergman, & Invine, 2007). Under this regime, the single body lumped temperature T evolution follows an Ordinary Differential Equation (ODE), Eq. (1), where A_{coo} stands for the external reference surface A_{ext} during cooling, currently the one that is in contact with air at temperature T_{at}. $(UA)_{coo}$ is the thermal conductance to the atmosphere.

$$C\frac{\mathrm{d}T}{\mathrm{d}t} = -\left(UA\right)_{coo}\left(T - T_{at}\right) \tag{1}$$

One problem with the single lumped thermal mass is that isothermicity is seldom reached exactly. To improve the applicability of this model, Eq. (2) indicates how to evaluate the steady-state heat capacity of the whole body C. If one considers that a body of average temperature T_k and heat capacity C_k evolves along another of average temperature T_j and heat capacity C_j, a heat balance between them only considering the heat accumulated is Eq. (2)[REMOVED REF FIELD]. In this equation, several bodies with different temperatures can be encompassed within the global heat capacity C at the average temperature T, considering that its corresponding variation in thermal energy is the sum of the thermal energies variations of such bodies.

$$C\mathrm{d}T = \sum_{1}^{n} C_j \mathrm{d}T_j \tag{2}$$

This allows incorporating into C parts that evolve with a different temperature. Eq. (3) indicates an easy procedure if $\mathrm{d}T_j/\mathrm{d}T$ can be evaluated against time.

$$C = \sum_{1}^{n} C_j \alpha_j \; ; \alpha_j \overset{\text{def}}{=} \frac{\mathrm{d}T_j}{\mathrm{d}T} \tag{3}$$

$j = 1, 2, \ldots n$ for the whole utensil and pot, formed by n bodies in thermal contact. This approach accepts known different temperatures during cooling to increase accuracy. As an example, the insulation, $\alpha_j = 1/2$ can be accepted, because a linear spatial inner temperature distribution from the water temperature T to the atmosphere can be presumed under quasi-steady conditions. Moreover, the plate temperature can differ from the pot content temperature, as is obvious in the result presented in the figures that follow, where a slight but almost proportional difference can be appreciated. After some analysis, here, $\alpha_j = 1 \; \forall j$ has been used because the insulation and plate heat capacities are marginal in comparison with the water heat capacity, and their values of α_j cannot be easily anticipated.

Defining the overtemperature θ, Eq. (4).

$$\theta \overset{\text{def}}{=} T - T_{at} \tag{4}$$

The ODE becomes as in Eq. (5).

$$C\frac{\mathrm{d}\theta}{\mathrm{d}t} = -(UA)_{coo}\,\theta + C\frac{\mathrm{d}T_{at}}{\mathrm{d}t} \tag{5}$$

Accepting that $\frac{\mathrm{d}T_{at}}{\mathrm{d}t} \ll \frac{\mathrm{d}T}{\mathrm{d}t}$, during a non-singular day one can accept Eq. (6).

$$\frac{\mathrm{d}\theta}{\theta} = -\frac{(UA)_{coo}}{C}\mathrm{d}t \tag{6}$$

From Eq. (6), Eq. (7) defines a characteristic cooling time t_{coo}.

$$t_{coo} \overset{\text{def}}{=} \frac{C}{(UA)_{coo}} \tag{7}$$

The solution for $\theta\langle t\rangle$ with constant C and UA, and with initial conditions $t = t_{in}$ $;\theta = \theta_{in}$ is in Eq. (8).

$$\frac{\theta}{\theta_{in}} = \frac{T - T_{at}}{T_{in} - T_{at,in}} = \exp\left(-\frac{t - t_{in}}{t_{coo}}\right) \tag{8}$$

Knowing C, one can obtain the heat losses coefficient U during cooling, here using the external area A_{ext} as reference, solving for it in Eq. (7).

Given the value of C, one can determine the heat loss coefficient U during the cooling process by referencing to the external area A_{ext} and solving for it in Equation (6).

$$U_{coo} = \frac{C}{A_{ext} t_{coo}} \tag{9}$$

The characteristic time t_{coo}, can be easily obtained from experimental data. In particular, considering a utensil and pot shared temperature sample T_i for $i = 1, 2, 3, \ldots$ and atmospheric temperature $T_{at,i}$, or an average value, an average characteristic time starting at t_{in} can be obtained, Eq. (10). Solving for it in Eq. (8).

$$\overline{t_{coo,i}} = \frac{t_i - t_{in}}{\ln\left(\frac{T_{in} - T_{at,in}}{T_i - T_{at,i}}\right)} \tag{10}$$

Alternatively, an approach to an $\overline{t_{coo}}$ is to obtain a straight line minimum squared fit of the denominator of Eq. (10) against $t_i - t_{in}$. The slope of this line is a good statistical option, equivalent to consider valid the hypothesis of constant $\overline{t_{coo,i}}$. If not, the slope would be non-constant for different i. In a more straightforward way, using Eqs. (6) and (7) the algorithm for the instantaneous $t_{coo,i}$ can be, using a central differences finite time scheme, which is noise reducer, Eq. (11). It is a discretized version of the ODE in Eq. (6).

$$t_{coo,i} = -\frac{\theta_i}{\theta_{i+1} - \theta_{i-1}}\left(t_{i+1} - t_{i-1}\right) \tag{11}$$

Extracting an average value of $t_{coo,i}$ during a certain period, it would sensibly coincide with the last value of $\overline{t_{coo,i}}$ for the same period, Eq. (10).

If one considers a constant time increment between samples $\Delta t = t_{i+1} - t_i$, an instantaneous U_{coo} can be directly obtained using Eq. (6) and (9) without having to accept constant U_{coo}. As commented below Eq. (5), accepting as an approximation $T_{at,i+1} = T_{at,i-1}$ is the only necessary condition. The outcome is Eq. (12). Large noise can appear, as this scheme is of an instantaneous value. By inspecting the results, a statistical hypothesis can be applied.

$$U_{coo,i} = \frac{C}{A_{ext}}\frac{\theta_{i-1} - \theta_{i+1}}{2\Delta t\theta_i} \cong \frac{C}{A_{ext}}\frac{T_{i-1} - T_{i+1}}{2\Delta t\left(T_i - T_{at,i}\right)} \tag{12}$$

Obviously, $U_{coo,i}$ represents $U_{coo}\langle t\rangle 1$,a function of time. But, at the same time, one can consider that the driving force for thermal power exchange W_{Th} is $\left(T - T_{at}\right)\langle t\rangle$ under quasi-steady conditions (i. e. $W_{Th}\langle t\rangle = A_{ext}U_{coo}\langle t\rangle\left(T - T_{at}\right)\langle t\rangle$. The natural and forced convection as well as the radiation heat transfer coefficients h can be a function of $T - T_{at}$, although the low temperature differences can make them almost constant. This must be determined from the experimental results, on which instrumental uncertainties, meteorological variations, and non-steadiness can separate from the constant value.

In a simple way, the approach to theoretically estimate the quasi-steady-state U_{coo} can be considered as the result of heat transfer coefficients h through the existing paths from the liquid content of the pot to the atmosphere. One can assume only two paths, through the insulated area A_{ins} and through the non-insulated area A_c, typically, the lid of the pot, Figure 1. This way, the external area $A_{ext} = A_{ins} + A_c$. The heat resistance of the metallic pot can be neglected, and the water can be considered at a uniform temperature. The heat losses through the insulated path starts through the insulating jacket of thickness δ_{ins}, with heat conductivity k_{ins} and external area A_{ins} yielding $h_{ins} = k_{ins}/\delta_{ins}$. It continues, in series, with the in parallel mixed natural and forced average convection coefficients h_{fc} and h_{nc}, respectively, into the surrounding air with active area A_{ext}. A linearized average radiation heat transfer coefficient h_r must be added, also in parallel, to them. The mixed convection heat transfer can be approached using experimental correlations following well established guides, such as in (Incropera, DeWitt, Bergman, & Invine, 2007) among others. In Eq. (13) *Ra* and *Re* are respectively the Rayleigh and wind Reynolds numbers of the flow around the cooking utensil.

$$U_{coo,i}A_{ext} = \left[h_{ins}^{-1} + h_{ext}^{-1}\right]^{-1}A_{ins} + h_{ext}A_c \tag{13}$$

Eq. (14) explains the break-down of h_{ext}.

$$h_{ext} = \left(h_{fc}^3 + h_{nc}^3\right)^{\frac{1}{3}} + h_r \;;\begin{cases} h_{fc} \propto Re^{0.5} \\ h_{nc} \propto Ra^{0.25} \\ h_r \propto \sigma\varepsilon T_{ext}^3 \end{cases} \tag{14}$$

Despite its apparent simplicity, the theoretical calculation presented here is quite complex and is often influenced by uncertainties in practical scenarios. Here, an experimental estimation is followed. The results obtained during the test campaign indicate that a linear evolution of $t_{coo,i}$ is the highest order that can be resolved with the testing platform. Because of that, Table 1 indicates the value obtained at the start of the cooling process ($_{ini}$) and at its end ($_{end}$). It was discovered that during cooling, when the load temperature comes from a high initial value, like on May the 9th (see section 3.3), and it becomes less than about 5 to 10 °C above the ambient temperature, t_{coo} increases, almost duplicating the previous and consistent values, Table 1. The initial hypothesis to explain this phenomenon is that the heated external insulation gradually releases heat inward with a delay. This is due to its low conductivity. The heat delivered to the inside becomes comparable to the heat losses to the atmosphere at the end of the cooling process. Further research should test this hypothesis.

2.1.2. Heating Process

The application of a time varying PV power $W\langle t\rangle$ on the hot plate results in an uneven temperature distribution inside the utensil because of the concentrated power and the unavoidable finite thermal conductances inside the utensil, evident in Figures 4 to 7. The concentrated power is a paramount difference from the cooling process. With box cookers and parabolic concentrating solar cookers, the solar power is spread around the external surface of the pot. This would imply that during heating $U_{hea} \neq U_{coo}$, but U_{coo} can be considered a good estimation of U_{hea}. The uneven temperature would be minimized when the effective internal heat conductance from the PTCs to the water is sufficiently large in front of the losses heat conductance $A_{ext}U$, causing a low enough Biot number. Consequently, the temperature T would be primarily an average of the values from the PTCs to the water with some uncertainty. Considering these temperature differences negligible, a single lumped mass heat balance is offered at Eq. (15).

$$C\frac{\mathrm{d}T}{\mathrm{d}t} = W - (UA)_{hea}\left(T - T_{at}\right) - \dot{m}_{ev}L_{w} \tag{15}$$

The PTCs dissipated electrical power is $W\langle t\rangle = I\langle t\rangle V\langle t\rangle$. An ammeter and a voltmeter can measure respectively the intensity I and voltage V, connected to the PV panel output. As an alternative, power can be directly measured using a commercial power meter. The global heat transfer coefficient for the losses to atmosphere U_{hea} can be determined as in Eq. (16) by solving for it in Eq. (15).

$$U_{hea}\langle t\rangle = \frac{W - \dot{m}_{ev}L_{w} - C\frac{\mathrm{d}T}{\mathrm{d}t}}{A_{ext}\left(T - T_{at}\right)} \tag{16}$$

Only a model of several connected lumped heat capacities, or a correction to C, like in Eq. (3), would give a criterion on carefully selecting a representative T when heating. On the other hand, the evaporated water enthalpy power $\dot{m}_{ev} L_w$ requires on-line measuring the instantaneous evaporated mass flow rate $\dot{m}_{ev}\langle t \rangle$, what is not easy at all. If the pot is open to atmosphere, an approximation would be to make it proportional to the vapor pressure of water $p_w\langle T_w \rangle$. The proportional factor could be approximated before reaching boiling ($\frac{dT_w}{dt} \neq 0$) by measuring the total weight loss at the end. Because of these limitations, the algorithm in Eq. (10) or in Eq. (11) are herewith used alone, to determine t_{coo}. A comparison between both procedures would enhance confidence in the results.

2.2. Testing Procedure-and Derived Parameters

One purpose of the heating tests is oriented towards determining the power of the solar cooker for heating from ambient temperatures up to boiling water and to standardize it. A standardized procedure for that is (American Society of Agricultural and Biological Engineers, 2013) based on the work of (Funk, 2000). Other tests are for estimating U_{hea}, requiring an empty pot to reach the maximum, stagnant, temperature T_{ST} as an indication of the applied power over the high temperature losses power as in Eq. (15) with $\frac{dT}{dt} = \dot{m}_{ev} = 0$, so that, Eq. (17) gives its stagnant value.

$$U_{hea,ST} \stackrel{\text{def}}{=} \frac{W_{ST}}{A_{ext}\left(T_{ST} - T_{at}\right)} \tag{17}$$

In our case, this gives a check of the U_{coo} calculation but at a higher temperature than those of the initial cooling, when an acceptable isothermicity is reached.

Other tests emphasize the efficiency in using the solar energy. (Odoi-Yorke, Baah, & Opoku, 2025) offers a recent bibliographic orientation on the topic. (Lahkar & Samdarshi, 2010) offers a more specific guide. It is surprising that there is limited interest in evaluating the steady-state water evaporation capacity, likely because this process is highly energy-intensive, although it is essential for some recipes, and for frying. The topic of standard performance figures for solar cooking is not closes, so there is even a proposal for a new standard for characterizing solar cookers (Kundapur & Sudhir, 2009).

Efficiencies are always important figures of merit. Let us consider heating, either a load of water or oil. Thus, the thermal power W_{Th} is applied to a constant specific heat capacity $C_{load} = \left(m c_p\right)_{load}$. It accumulates from the initial temperature of the load inside the pot T_{ini}, usually near or at T_{at}, up to a temperature $T_{end} = 95$ (slightly less than the boiling temperature to avoid excessive vapor losses), forming the load thermal energy content E_{Th}. It can be considered that this accumulated thermal energy is fed by an energy input E_{input}, either solar E_s or after it is converted into electrical E_{PV} by the PV panel. The average solar efficiencies are defined in Eq. (18) following (Kundapur & Sudhir, 2009) and (Nahar, 2003) among others.

$$\eta_{Th,95} = \frac{E_{Th}}{E_{input}} = \frac{(mc_p)_{load}(95 - T_{ini})}{E_{input}} \tag{18}$$

If the global solar efficiency is considered, one has the solar efficiency $\eta_{Th,s,95}$. Eq. (19) details $E_{input,\ s}$, using samples separated $\Delta t = t_{i+1} - t_i \forall i$. In this expression, $\overline{G_{T,i}}$ is the average during Δt.

$$E_{input,\ s} = A_a \int_{T_{ini}}^{95} G_T \mathrm{d}t \cong A_a \Delta t \sum_{i_{ini}}^{i_{95}} \overline{G_{T,i}} \tag{19}$$

If the electrical efficiency is considered as input, $\eta_{Th,PV,95}$ evaluates the thermal part of the utensil, thermal efficiency, thus excluding the electrical part. Eq. (20) details $E_{input,\ PV}$.

$$E_{input,\ PV} = A_a \int_{T_{ini}}^{95} W \mathrm{d}t \cong A_a \Delta t \sum_{i_{ini}}^{i_{95}} V_i I_i \tag{20}$$

An instantaneous utensil heating efficiency can also be considered for the instruments sample i Eq. (21). Using a central differences algorithm for $\eta_{Th,PV,i}$.

$$\eta_{Th,PV,i} = \frac{(mc_p)_{load}(T_{i+1} - T_{i-1})/(2\Delta t)}{W_i} \tag{21}$$

The electrical part is the result of the PV panel and the dissipation circuitry, including the PTCs. The combination of all that configures the here called PV panel instantaneous efficiency $\eta_{PV,i}$, herewith considered under the same prior principles, Eq. (22).

$$\eta_{PV,i} = \frac{W_i}{A_a G_{T,i}} \tag{22}$$

If this efficiency is averaged from the start, up to $T = 95$, it becomes $\eta_{PV,95}$.

The overall solar instantaneous (Δt average) efficiency is $\eta_{Th,s,i} = \eta_{Th,PV,i}\eta_{PV,i}$ .and if it is averaged the from start, up to $T = 95$ is $\eta_{Th,s,95}$ or $\eta_{Th,PV,95}$, according to Eqs. (18) to (20).

When oil is loaded into the pot, it can withstand higher temperatures, allowing for the maximum (stagnant) temperature T_{ST} to be recorded with minimal vapor losses. However, some oxidation or degradation of the oil may occur, changing its thermal properties. The inconvenience is that there is a temperature gradient in the thermal part of the utensil, from PTCs to the oil-containing pot. Consequently, the figure for losses obtained is not strictly accurate in terms of a single lumped thermal mass. If a cooling process without any applied power follows, after a satisfactory temperature homogeneity is reached, evaluating heat losses can be performed with acceptable accuracy.

During heating, some researchers, (Funk, 2000) and followers, advocate for $T_{end} = T_{ini} + 50$, especially for solar cookers that are not intended for water boiling, as cooking and many pathogens destroying are attained when reaching 70 °C for some time. This alternative standard gives different figures of efficiency, η_{50}. In this test, the previous power delivered should be quasi-constant, according to an accepted rating (Funk, 2000). The standard heating power W_{st} is obtained with a correction transforming the power to a solar irradiance normal to the aperture area of $G_T = 700 \text{ W m}^{-2}$ as Eq. (23) indicates 2.

$$W_{s,50,st} = W_{50}\frac{700 \text{ W m}^{-2}}{G_{T,50}} \tag{23}$$

Of interest in our case is also $W_{PV,50,st}$, and $W_{Th,50,st}$, correcting the actual powers like in Eq. (23).

Alternatively, or complementary, $T_{end} = 95$ configures the boiling time t_B. It can be attained using constant power or the natural varying power of the sun on a representative day, e. g. during the morning. In this case, specifying the operating conditions and obtaining an average $\overline{G_T}$, for the whole heating interval. (Lahkar & Samdarshi, 2010) proposed normalizing it to $G_T = 900 \text{ W m}^{-2}$ to obtain $t_{B,st}$ as in Eq. (24) and to use the standard load mass.

$$t_{B,st} = t_B\frac{\overline{G_T}}{900 \text{ W m}^{-2}} \tag{24}$$

Due to the innovative design of the mPVSC, two standard performance parameters are widely used, e. g. (Mullick, Kandpal, & Saxsena, 1987) but are considered meaningless here. This is because their value varies widely along the operation of the cooker and its non-linear response. Thus, they do not uniquely characterize its performance. These two parameters, generally known as F_1 and F_2, have been use in the past by the authors for other cookers types, e. g. (Lecuona-Neumann, 2017) and are adopted by the Indian standard (Bureau of Indian Standards (BIS), 1992) and (Bureau of Indian Standards (BIS), 2000).

An example of the rationale under which they become meaningless is that $F_1 = (T_{ST} - T_{at})/G_T$, is based on the empty pot maximum temperature T_{ST}, associated to G_T. A large value of T_{ST}, associated with superior performance, would never be used for real cooking (for safety reasons and because the pot would not be empty under normal operation). The mPVSC is conceived with a built-in temperature limit slightly higher than T_C (around 220 °C in our case). This enhances safety and robustness. In an empty pot (except for very low values of G_T), the mPVSC T_{ST} would be independent of G_T. As the PTCs electrical resistance increase is the limiting factor, it would not characterize the performance of a filled pot. The formulation for F_2 is more complex, but the rationale is the same. Section 3.3 further illustrates this point as it reports oil heating up to T_{ST}. In that test, F_1 can be considered comparable to that of a conventional empty pot. The standard does not use filled cases because of the longer time needed to reach stagnation and the use of hot oil capable of spontaneous burning.

To establish a foundation for testing and due to the innovative design of the mPVSC, heating and subsequent cooling tests have been conducted under various conditions to assess more accurately the capabilities of the present implementation of an mPVSC. Due to the variations among the tests, the most pertinent and easily understandable figures of merit are presented along the paper. The spirit and aim of the different standards are respected, but there is no intention in this paper to fulfill a standard rating for the mPVSC.

2.3. Experimental Setup

Figures 1, 2, and 3 depict the testing platform, which was used outdoors for a good representation of its practical use, under non-ideal conditions. It includes:

- A 72-cell solar polycrystalline panel, REDSOLAR RED310-72P, of W_{max} = 310 W_p, V_{MPP} = 36.5 V, and V_{max} = 46.0 V at ambient temperature. The aperture area A_a was 195.6x99.2 cm. It was oriented to the south and with a 45 deg fixed tiling angle. The location coordinates were 40° 19' 37.85" N, 3° 45' 48.60" W.
- An mPVSC utensil, which is supported by a wooden base, Figure 2. The operating console includes an in-built digital consumer-grade power meter of generic Chinese origin. A main switch, Sw0, allows applying power or diverting it externally. Three PTC switches allow power maximization by connecting them in parallel, Figure 1(d). This design is in accordance to Lecuona-Neumann et al. (2024).
- A brass hot plate of 12.4 cm diameter and 5 mm thick. It is insulated at its bottom by a 4 mm thick cork plate. Three socketed flat plate commer-

cial PTCs heating elements of 60 mm×21 mm, of generic Chinese origin, are equally spaced on its underside and fixed by screws. They are originally sandwiched using a Kapton® plastic insulating socket, enclosing two contact plates and soldered terminals. This commercial set is presented originally pressed inside an external aluminum coaxial flat socket. They individually exhibit a minimum resistance of about 6 to 8 Ω close to T_C = 200 to 220 , Curie temperature, showing an exponential growth at higher temperatures. At ambient temperature, their individual resistance was around 40 Ω.

- A cylindrical aluminum pot (red on the figures) of 15 cm diameter, 7 cm tall, and of 1.5 mm thick walls. A flat removable lid, made of 0.5 mm thick stainless-steel sheet, completes the cooker food compartment.
- A moveable rock wool insulating lateral jacket. It conforms a hollow cylinder with inner and outer diameters of 18 cm and 25 cm respectively, and a height of 14 cm. It is covered by a thin aluminum adhesive foil, Figure 2. Its installation increases A_{ext} of the pot plus the hot plate and PTCs set. Additional thermal insulation was devised for testing with the pot containing oil and to demonstrate the possibility of reaching high temperatures during heating and cooling without damage. This increased insulation was formed by old cotton towels, Figure 3.
- 7 Type K, Class A, bare head thermocouples connected to a Pico technology® TC-08 A/D converter, linked via a USB cable to a portable PC, running the PicoLog® data acquisition software. They provided average measurements for each Δt during the measurement campaign. They measured at the following points: two of them the underside of the hot plate, another two measured the water temperature, another measured the side of the pot or the external surface of the insulating jacket. An extra one was fixed to the top pot lid. The atmospheric temperature was measured by a PT100 thermometer, certified at a standard deviation $\sigma = 0.5$ as the random error uncertainty. The total measuring uncertainty of the thermocouples when installed was estimated at $\sigma = 1$ despite calibrating the thermocouples to a lower uncertainty. As the temperatures change during the time interval Δt, even averaging measurements each 10 minutes, there is still an uncertainty component remaining. The panel temperature was measured at its blackened surface with an IR thermometer, giving a total uncertainty of $\sigma = 1.5$.
- External industrial electrical instruments. A Hall-effect digital ammeter for measuring I and a multimeter for measuring V. Both worked as a check of the in-built power meter to correct its less-than-ideal systematic error of +15%. These external meters were used as calibrators owing to their much smaller nominal total uncertainty, in the order of $\sigma = 1\%$. The expanded total uncertainty (95% probability) of the power measurement $W = VI$ was judged to be

$2\sigma = 10\%$, including the meters uncertainties and the time variations during the period of measurements Δt plus the reading uncertainty. The working PTCs' resistance was obtained as $R = V/I$ with an expanded total uncertainty of $2\sigma = 10\%$. Both V and I were considered independent measurands.

- An anemometer for wind speed. During the experiments, it remained in the interval of 0 to 2 m/s from a random direction. This makes the wind effect within the uncertainty of the measurement chain, and it is within the allowance of the standards.
- A semiconductor based MacSolar® global solar meter oriented parallel to the PV panel plane. It measured the tilted irradiance G_T with a total uncertainty of $\sigma = 10$ W m^{-2}.

Uncertainties in C and A_{ext} were judged as negligible in front of the uncertainty in t_{coo}, which was estimated by the dispersion on the experimental data used for its assesment. Following the methodology in (Holman, 2011) and (ISO, 1995), the collection of the instruments uncertainties on place gave an expanded uncertainty in t_{coo} and U_{coo} as two times the standard deviation (2σ), either around the average or around the least square straight line, plus 10% of the expanded uncertainty of the composition of the instruments, according to Eqs. (7) and (9).

3. EXPERIMENTAL RESULTS

All the tests were conducted in the same location in May 2023, utilizing the same PV panel, utensil, pot, and instruments. In the figures, the time is displayed as hours and their decimal data. All the tests were performed measuring every $\Delta t = 10$ minutes.

As a general issue, there are two main possibilities when heating starts: (i) When irradiance is low, such as in the morning, utilizing fewer than three PTCs (-1- or -2-) is optimal for maximizing power, as indicated in the previous paper Lecuona-Neumann et al., (2024). (ii) Under high irradiance conditions, such as when starting near midday, the three resistances maximize power, especially if the PTCs are still hot because of a previous use, or after a PCTs heating transient. Also, during sunset or cloudy periods, it may be advantageous to operate with fewer than three PTCs in parallel connection.

Regarding the structure of the measurement campaign, in May 2025, for this paper:

- On the 10th, after the heating process was completed, the utensil was briefly set aside for inspection. Cooling was then initiated and monitored closely. Two consecutive experiments (heat plus cooling cycles) were conducted on this same day: the first in the morning with an uninsulated (bare) pot, and the second one in the afternoon with the jacket thermal insulation installed, Figure 2. The morning was clear and sunny, while the afternoon was less favorable.

- On the 25th, a continuous test of the heating and cooling processes using the jacket-insulated pot filled with water was conducted throughout the entire day.
- On the 9th, 0.436 kg of anhydrous virgin glycerol oil was introduced into the pot to explore higher temperatures while simultaneously avoiding evaporation losses.

The results are presented in the following sections, beginning with water and subsequently addressing oil contents.

3.1. Uninsulated Pot Testing on May 10th, 2023

3.1.1 Heating During the Clear and Sunny Morning

The utensil with the bare pot closed by the drop-in lid was assessed first, starting the morning of May the 10th to simulate cooking for an early lunch.

Figure 4 resumes the results. Heating started at 9:55 hr local official time. Heating finished at 12:55 hr, before solar noon, when some vapor was detected coming from under the lid, so that the boiling time was $t_B = 2.5$ hr, corresponding to a standardized $t_{B,st} = 2.07$ hr. The cooling period was from 13:05 until 14:15, near solar noon, thus leaving time for later tests. Precise isothermicity was not possible in the cooling experiment owing to the non-common path of heat losses toward the ambience as different insulation was operating on each side of the utensil, top and bottom, and there was some internal heat resistance. As Figure 4 indicates, the cooling curves run parallel so that the temperature differences do not affect substantially t_{coo}. After the cooling process, the weight of the water was measured at 0.960 kg, reflecting an evaporation loss of 40 g. This value was used in the calculations for the cooling process. The PV power curve shows some peaks followed by ~10% power drops. These power drops correspond to situations where the load forces a transient higher voltage than the V_{MPP} because of the operator introduces an extra PTC to see if there is a power increase, This PTC is at is at a lower temperature than when electrically loaded for some time. For the gaps with lower voltage than the V_{MPP}, a few minutes later the power increases proportionally to the solar irradiance. This condition is the one seek by the operator of the cooker switches. The three PTC resistances stabilized between 7 and 8 Ω, while the maximum solar irradiance $G_T = 1$ kW m ² at the conclusion of the heating period. To reach MPP the three

PTCs loading should reach a nominal value $R_{MPP} = V_{MPP} I^{-1}_{MPP} = 36.5\ \text{V}/8.49\ \text{A} = 4.3\ \Omega$ plus the wiring resistance, which was not attained in this test and on none of the other ones. The instantaneous (during period Δt) $\eta_{PV,i}$ decreased from a starting value around 12% down to 8.7% at the end of heating. This suggests that the PTC set loading resistance is too high to approach around 18.5% of the PV panel's nominal efficiency at working temperatures. More and/or larger size PTCs seem advisable. Figure 7 also suggests this issue.

Although the power delivered was non-quasi-constant, as indicate in the standards, focus can be put on the point with 50 °C of overtemperature ($\theta_{end} = (T - T_{at})_{end} = 50$). At this point, the PV power was $W_{PV,50} = 159$ W, which according to Eq. (23) resulted in $W_{PV,50,st} = 143$ W, Table 1. The water heating power was $W_{Th,50} = 36$ W.

Boiling was reached after $t_B = 2.5$ hr, which resulted in $t_{B,st} = 2.07$ hr, according to Eq. (24), Table 1; yielding $W_{PV,95} = 170$ W under $G_T = 960\ \text{W m}^{-2}$. The average PV solar efficiency up to t_B was $\eta_{Th,s,95} = 2.48\%$, while the average utensil thermal efficiency was $\eta_{Th,PV,95} = 25.1\%$.

During the cooling process, using the water temperature, the characteristic cooling time for the first measurement Δt resulted practically the same as averaging up to the end of the cooling process $\overline{t_{coo}}$. Averaging from 13:13 hr local official time (3.3 hours from the start) and applying Eq. (11). It was found to be $t_{coo} = 1.49$ hr on average, Table 1. Between the end of heating and start of the cooling process, there appeared a significant heating of the lid, almost reaching the pot lateral temperature, as shown in Figure 4. This suggests a significant two-phase heat transfer between the evaporating water surface and the lid condensing water, because of the water was almost boiling.

The total surface area of the utensil was assumed to be a closed cylinder from bottom to top, with an external area of $A_{ext} = 0.0683\ \text{m}^2$. This datum and the computed entire utensil heat capacity $C = 4.466\ \text{kJ K}^{-1}$, allowed to obtain an average $U_{coo} = 16\ \text{W m}^{-2}\ \text{K}$, applying Eq. (9), as indicated in Table 1. This datum is consistent with correlations for convection and radiation heat losses of a vertical plate or a cylindrical surface closed by an upward and downward horizontal plate for respectively the lid and bottom surfaces, Eq. (13), using the theory in (Rohsenow, Harnett, & Cho, 1998). The expanded uncertainty around this average value during the cooling process was $2\sigma = 27\%$, which, according to the previously estimated instruments uncertainty, gives a mean squared total expanded uncertainty of $2\sigma = 29\%$.

3.1.2. Heating During the Afternoon With Partial Cloud Cover.

On the same day, May 10th, the empty pot was refilled with 1 kg of water at ambient temperature and closed with the lid once again. It was then placed on the hot plate and laterally bordered with the insulating jacket closing at its bottom against the supporting base, Figure 2. At 14:30 hr, the test began with the three PTCs connected. Figure 5 illustrates the heating and cooling monitoring results, starting at 14:35 hr official time and ending the heating at 16:55 hr. The afternoon was less sunny than the morning, as clouds drifted by. This is evident in Figure 5 with drops in G_T. They were accompanied by transient drops in power and panel temperature. During this test, water did not boil, so the loss of weight was negligible.

The maximum efficiencies were: $\eta_{Th,PV,i}$ = 45% at 14:55 hr, corresponding to 14.92 hr in decimal notation, while $\eta_{Th,s,i}$ = 2.6% 30 minutes later, and $\eta_{PV,i}$ = 16% 40 minutes later.

Figure 5 shows the lowest resistance reached just at the start with – 3 - PTCs connected, but after the first measurement, 10 minutes after the start, there was a plate temperature higher than T_C, reaching 229 °C, while water was at 48 °C, highlighting the temperature differences addressed before. It is likely that the insulation jacket, which laterally covers the plate and PTCs, is responsible for providing enhanced insulation, which caused this high plate temperature. This is helped by residual heat on the hot plate. As $T > T_C$, the load electrical resistance is higher than R_C and higher than R_{MPP}, limiting power. Both standard powers, PV and thermal, using Eq. (23) resulted in smaller values than during the morning because of too high PTCs resistance, Table 1. This phenomenon underpins two issues. One, assuming a single lumped temperature during heating is inaccurate for a precise modeling of the present design during heating as simultaneously the PTC are above T_C and the water is not boiling. Two, a better thermal conductance between PTCs/plate and even plate/pot are needed to reduce the temperature difference between PTCs and water during heating. This calls for eliminating both sockets that envelop each commercial PTC, thus using bare PTCs elements instead of them, Figure 1(b). This would produce direct electrical and thermal contact with the hot plate. In our case, this would require three separate flat electrodes compressing each PTC against the hot plate but electrically insulated between them. The use of conductive grease or cement seems convenient to further reduce the contact resistance and add thermo-mechanical compliance. A larger hot plate and pot diameters, jointly with a flat and rigid pot bottom, would increase thermal conductance to water.

The cooling showed to be slower than with the bare pot. Applying Eq. (10) resulted in some variation along the cooling process. Thus, Eq. (11) seems more appropriate than Eq. (10) to this test. When applied to water temperature after the approximate isothermicity results in $t_{coo,ini}$ = 2.4 hr linearly evolving to $t_{coo,end}$ = 2.6 hr with an expanded uncertainty on the data of 2σ = 3%. Adding the instruments uncertainty, this yields a final expanded uncertainty of 2σ = 11%. With C = 4.533 kJ K^{-1} and A_{ext} = 0.208 m^2, results in U_{coo} = 2.4 W m^{-2} K^{-1}, 83% less than with the bare pot, but with a larger A_{ext}. As a reference, this is around twice the nominal heat conduction radial insulating h_{ins} of the thickness of the insulation jacket alone, Eq. (13), what seems coherent considering the additional heat losses of the air flow short-circuits and the non-insulated lid, Figure 2.

3.2. Continuous Whole-Day Test on May 25th, 2023

The 25th of May was a bright and clear day. Figure 6 illustrates the results. Heating commenced at 10:57 hr, which was later than the morning heating sessions of the previous 10th. All the three PCTs (-3-) were connected from the start making the lowest electrical resistance of all the tests, even from the start. Before noon, at 12:17 hr, the heating was halted once the boiling temperature t_B was achieved after 1 hour and 20 minutes (1.33 hr). Monitoring continued uninterrupted during the cooling phase. At 14:00 hr, the water temperature had cooled down to 60 °C, marking the point at which cooking was considered complete, EoC. Thus, in-place cooking endured 3 hr. The water exhibited again a loss of 40 g.

The plate temperature did not reach T_C. The PV panel efficiency was maintained at about $\eta_{PV,i} \cong 9\%$ during the whole heating process, far from MPP. The maximum instantaneous thermal efficiencies were $\eta_{Th,PV,i} = 66\%$, a respectable value, and $\eta_{Th,s,i} = 6.3\%$ at 11:37 hr, corresponding to 11.62 hr in the decimal notation used in the figures.

The analysis following Equation (10) revealed an average cooling characteristic time of $t_{coo} = 2.8$ hr, with an expanded total uncertainty of $2\sigma = 44\%$. From start util t_B, the average efficiencies became: for the solar panel $\eta_{PV,95} = 9.2\%$ and for the overal (usensil+pot) $\eta_{Th,PV,95} = 51\%$ giving a global solar efficiency $\eta_{Th,s,95} = 4.7\%$.

Considering a representative convection plus radiation coefficient of only the external lid surface $h_{ext,lid} = 10$ W m^{-2} K^{-1}, the power loss at the end of heating was around 8.8% of the PV power of 153 W. This highlights the importance of a complete thermal insulation during heating, including the lid. The next section addresses this issue.

3.3. Continuous All-Day Test on May 9th, 2023 Using a Highly Insulated Utensil and Containing Oil

This test on May 9th corresponds to the configuration of Figure 3. As Figure 7 indicates, a stagnation plate temperature of 235 °C > T_C was reached, just before stopping power under also a stagnant G_T at 4.7 hours after start. The almost constant and near-atmospheric temperature of the insulating jacket's external lateral temperature indicates a good lateral and bottom thermal insulation by the jacket. The maximum instantaneous solar panel efficiency $\eta_{PV,i} = 12.4\%$ at the same time as the maximum overall efficiency $\eta_{Th,s,i} = 6.1\%$, at 09:57 corresponding to 9.95

hr,. This is 20 minutes before reaching the maximum utensil thermal efficiency of $\eta_{Th,PV,i} = 57\%$.

For calculating $t_{coo,i}$, data were accepted starting at 14:00 hr. Because of the large overtemperature reached, the difference between initial and final $t_{coo,i}$ were the largest. At the end of the cooling process $t_{coo,i}$ was the largest of the testing campaign, Table 1. At the start of cooling, its value was not larger than with less insulation, in principle as a result of the large overtemperature. Because of the substantial mass of the insulating towels an appreciable heat is given back by them at the end of the cooling process, which is not taken into account in the single lumped thermal mass model.

The PV voltage remained relatively stable at $V > V_{MPP}$ (nominal) after an initial period, even with the addition of PTCs during the heating process. The lid temperature was approximately 12 °C lower than the water temperature, which was not a smaller difference compared to the other tests, despite the use of a high thermal insulation on the upper side, additionally, because oil was used instead of water. This implies that the heat transfer associated with water evaporation from the liquid evaporating surface to the condensation at the lid was absent.

Applying Eq. (17) for the stagnation, reached 14:09 hr, resulted in a thermal losses conductance $U_{hea,ST}A_{ext} = 0.33$ W K^{-1}. Even counting with the power limiting effect of the PTCs because $T > T_C$ when stagnation is reached at $T_{ST} = 235$, the first figure of merit F_1, according to (Mullick, Kandpal, & Saxsena, 1987) reached a value $F_1 = 0.19$ clearly superior of the stipulated minimum value of 0.12. Assuming an estimated $A_{ext} = 0.3$ m^2, according to Figure 4, results in $U_{hea,ST} = 1.1$ W m^{-2} K^{-1} . This value is in contrast with the values obtained during the cooling process, according to Table 1, $U_{coo,ini} = 0.60$ W m^{-2} K^{-1}, and $U_{coo,end} = 0.28$ W m^{-2} K^{-1}, with an average value of 0.36 W m^{-2} K^{-1} although at lower temperatures and having reached an approximate isothermicity. On one hand, these non-coherent values highlight the importance of the isothermicity hypothesis for the single lumped thermal capacity model. On the other hand, these values are significantly lower than the other days' values, Table 1. This can be explained in two coinciding circumstances, (i) a much larger A_{ext} and (ii) a much smaller overtemperature of the external towel surface. The latter makes $Ra \rightarrow 0$ making h_{cn} of small value, Eq. (14). h_r can also be small as well; this is initially attributed to a radiant environment of temperature higher than both the sky and atmospheric temperatures, as the testing platform was located beneath the PV panel, where temperatures exceed those of the atmosphere, especially during the cooling process, when $W = 0$. The utensil also "views" the floor, interchanging thermal radiation with it, whose temperature is significantly higher than the surrounding air temperature when solar irradiance is high enough, which was the case. Thus, the external towels' temperatures can be almost equal to the environment´ average radiant temperatures.

3.4. Analysis of the Overall Results

Figure 8 shows the resulting voltage because of the operator's action for all the tests against time, after starting. Observing this Figure 8, it evidences that during the whole testing campaign, the voltage is near V_{MPP}, except during the initial stages of heating when the utensil is cold and/or there is low irradiance, as fewer than three PTCs were connected, all this originating $V < V_{MPP}$. This did not happen during the reheating on the afternoon of May the 10th, as there was residual heat inside the hot plate and simultaneously a high irradiation. $V > V_{MPP}$ towards the end of heating happened for all the tests excepting one, during the morning of May the 10th because of an imperfect operator action. $V > V_{MPP}$ suggests that the PTCs area is not enough, or equivalently, a too large overall electrical resistance.

From the experience gained, one can conclude that a power meter could be replaced with a simple voltmeter in favor of simplicity and low cost, even an ordinary external one. This requires only two contacts instead of four. In this situation, the operator would select a specified voltage $V \cong V_{MPP}$, unless her/he determines that the irradiance is low, in which case a lower voltage is accepted. Furthermore, in the simplest form of mPVSC, all the PTCs can be connected to the panel, which may lead to some power loss, particularly during periods of low irradiance and especially when water temperature is low, which can be qualitatively sensed without a thermometer. In this scenario (all PTCs connected), no operator judgment or actions are required, aside from simply switching the system on or off. Another possibility is that the switches can be replaced by electrically insulated screw connections of the cables, less prone to failure than switches as they switch a high-power DC. This possibility would result in lower cost but leading to mistakes. The lower than 40 V voltage would make this harmless in case of an electrical shock.

The shortest heating time was recorded on the 25th, when the insulating jacket was sealed against air short-circuiting between the pot lid and the jacket, coupled with high irradiation levels. But Figure 8 also indicates that on the 25th, the hot plate did not reach T_C. This is attributed to a less than perfect sealing in the PTC zone, below the hot plate, owing to the removable character of the insulating jacket. It is presumed that a more tight jacket would approach T_C more easily and, as a consequence, provides more power, as a result of lower R. Only by facilitating the heat transfer PTCs/food can an excessive PTC temperature be avoided, in order to avoid limitations for frying. When oil is heated on the 09th, Figure 7 indicates longer heating times as the reached temperatures were higher, but 95 °C were reached after 1.5 hours of starting time, one of the shortest times, although the heat capacity in this case C was the lowest.

Table 1. Derived parameters from results of the different days.

Day of May 2023	Description	$W_{PV,50,st}$ [W]	$W_{Th,50,st}$ [W]	$t_{B,st}$ [hr]*	$t_{coo,ini}$ $\pm 2\sigma$ [hr]*	$t_{coo,end}$ $\pm 2\sigma$ [hr]*	$U_{coo} \pm 2\sigma$ $\left[\frac{W}{m^2 K^{-1}}\right]$**
10th	Morning	143	32	2.07	1.49±14%	1.49	16
10th	Afternoon reheating with jacket	60	16	n. a.	2.4±5%	2.6	2.4
25th	Late morning with jacket	143	32	2.07	2.8±44%	2.8	2.4
09th	Oil, high insulation	133	74	1.00**	2.8 ±11%	5.8	0.36***

*: Decimal. ** Virtual between start and end, referred to A_{ext} . ***: Lower C and non precise large A_{ext}.

4. CONCLUSIONS

This paper offers an experimental evaluation of a minimal innovative Photo-Voltaic Solar Cooker mPVSC. It enables indoor cooking and electricity supply. It functions without electronics or batteries. This design offers very simple and most affordable way by directly connecting the DC from PV panels to a set of Positive Temperature Coefficient (PTCs) commercial resistances through a set of switches and a power indicator.

The findings from the experiments on a mPVSC demonstration unit can be summarized as follows.

- It becomes feasible to eliminate the reliance on electronics or batteries for PVSCs. This approach not only enhances robustness but also reduces costs and eludes fire hazards. Without batteries, cooking with the help of Thermal Energy Storage (TES) allows for the possibility of preparing meals when the sun is not shining.
- A power meter provides users with guidance on how to operate the PTC switches for optimal power management. Fortunately, this operation is not frequently required. Simple judgement can replace the power meter, e. g. when there is a significant decrease in solar irradiance and the load is cold, it is convenient to simply observe the sun and switch to fewer PTCs. For high irradiance levels, all three switches should be closed excepting with a cold pot, just for some time. Even more, a design without switches and power meter is also feasible. This design, in addition to no fire risk, enables fully unassisted solar cooking, although a fuse is recommended.

- The data obtained suggests that the design implementation here reported could enable a greater number of PTCs to enhance power extraction from the PV panel, as theoretically noted in a previous article.
- The heating experiments demonstrate that boiling 1 kg of water can be achieved in a reasonable timeframe during the morning, even when utilizing a bare, non-insulated pot and insufficient PTC area. A second and even a third boiling can be conducted following the initial one on a clear, sunny day.
- Using an insulation jacket allows for quicker boiling and especially better temperature retention, which extends cooking time and enables meals to be enjoyed later in the afternoon.

The findings enable decision-makers to incorporate the mPVSC concept as a tool in the fight against energy poverty and vulnerability. They can assist engineers in predicting the performance of PV solar cookers, thereby streamlining the design process. For conventional thermal cookers utilizing the thermal loss conductance data obtained from the cooling experiments conducted at various levels of thermal insulation can be useful in the design process. The data indicates that a single lumped thermal mass model for characterizing solar cookers is limited, especially during heating.

When scaling this technology to larger sizes, community mPVSC cookers are feasible, such as for schools, hospitals, and community centers.

As future works, factors such as hygiene, daily usage, and long-term durability necessitate comprehensive research to guarantee practicality and sustainability in real-world applications. The hypotheses raised on explaining the thermal flow inside the utensil need further investigation.

The concept of mPVSC can be applied to configure a minimal PV Electric Pressure Cooker (EPC) mimicking both the AC-driven commercial ones, e. g. (Ossei-Bremanga & Al, 2024), and the DC-driven but including electronics and batteries as described in (Kinyua, Wambugu, & da Silva, 2025). An electric pressure cooker is a versatile cookware that uses steam and pressure to cook food quickly and efficiently, including thermal insulation. It operates by sealing in steam, which raises the internal pressure and temperature, allowing food to cook faster than traditional methods. EPCs are ideal for preparing a variety of dishes, including soups, stews, grains, and even fried and sauté recipes. Numerous models feature programmable settings that enable users to customize cooking times and temperatures for a variety of recipes.

CONTRIBUTION STATEMENT

A .L.: Writing – original draft, review & editing, investigation, formal analysis, conceptualization, platform design and construction, experiment assistance.

J. B: Methodology, conceptualization, design and utensil manufacturing, investigation.

J. I. N.: writing, review & editing, formal analysis, data curation, administration, instrument provision.

MC. R. Writing, review & editing, data curation, experiments assistance.

ACKNOWLEDGMENT

The authors would like to express their sincere gratitude to the technicians of the Departamento de Ingeniería Térmica y de Fluidos at the Universidad Carlos III de Madrid: Manuel Santos, David Díaz, and Israel Pina. They also wish to acknowledge the valuable contributions of student Jaime Martínez-Barquero, who assisted in conducting experiments as part of his Master thesis.

REFERENCES

Aemro, Y., Moura, P., & de Almeida, A. (2021). Inefficient cooking systems a challenge for sustainable development: A case of rural areas of Sub-Saharan Africa. *Environment, Development and Sustainability*, *23*, 14697. https://link.springer.com/article/10.1007/s10668-021-01266-7

American Society of Agricultural and Biological Engineers. (2013). *ASAE Standard S-580.1 Testing and Reporting Solar Cooker Performance.* Standard, ASABE St. Joseph, Michigan, USA.

Arunachala, U., & Kundapur, A. (2020). Cost-effective solar cookers: A global review. *Solar Energy*, *207*, 903–916. DOI: 10.1016/j.solener.2020.07.026

Atmane, I., Kassmi, V., Deblecker, V., & Bachiri, V. (2021). Realization of autonomous heating plates operating with photovoltaic energy and solar batteries. *Materials Today: Proceedings*, *45*(8), 7408–7414. DOI: 10.1016/j.matpr.2021.01.604

Bailis, R., Drigo, R. A. G., & Masera, O. (2015). The carbon footprint of traditional. *Nature Climate Change*, *5*, 266–272.

Bruce, N., Perez-Padilla, R., & Albalak, R. (2000). Indoor air pollution in developing countries:a major environmental and public health challenge. Bulletin of the World Health Organization, *78*(9), 1078. Retrieved 01 08, 2024, from https://pubmed.ncbi.nlm.nih.gov/11019457/

Bureau of Indian Standards (BIS). (1992). *Indian Standard-Solar Cooker, IS 13429.* Manak Bhawan.

Bureau of Indian Standards (BIS). (2000). *Indian standards IS 13429: solar cooker e box type, first revision.* BIS, Manak Bhawan.

Cardoso, J., Couto, A., Costa, P., Rodrigues, C., Facão, J., Loureiro, D., & Simões, T. (2023). Solar Resource and Energy Demand for Autonomous Solar Cooking Photovoltaic Systems in Kenya and Rwanda. *Solar, 3*, 487-503. https://www.sciencedirect.com/science/article/pii/S0038092X21002280

Das, I. a. (2025). The Costs and Benefits of Clean Cooking Policies in Low- and Middle-Income Countries Under Real-World Conditions. *Sustainable Development, n. a.*(n. a.), n. a. DOI: 10.1002/sd.3463

De, D., Shawhatsu, N., De, N., & Ajaero, M. (2013). Energy-efficient cooking methods. *Energy Efficiency*, *6*, 163–175. https://link.springer.com/article/10.1007/s12053-012-9173-7

Funk, P. A. (2000). Evaluating the international standard procedure for testing solar cookers and reporting performance. *Solar Energy*, *68*(1), 1–7.

Geng, X., & Bai, L. (2023, October). Characteristics of particulate matter and polycyclic aromatic hydrocarbon pollution generated during kitchen cooking and health risk assessment. *Indoor and Built Environmen, n.a.*(n.a.), n.a. DOI: 10.1177/1420326X231219999

Harmim, A., Merzouk, M., & Boukar, M. M. (2013). Design and experimental testing of an innovative building-integrated box type solar cooker. *Solar Energy*, *98*(C), 422–433. DOI: 10.1016/j.solener.2013.09.019

Holman, J. (2011). *Experimental methods for engineers* (18th ed.). McGraw-Hill Education.

Incropera, F., DeWitt, D., Bergman, T., & Invine, A. (2007). *Fundamentals of Heat and Mass Transfer* (6th ed.). John Wiley.

ISO. (1995). *BIPM, IEC, IFCC, ISO, IUPAC, IUPAP, OIML. Guide to the Expression of Uncertainty in Measurement*. International Standards Office.

Kaushik, S., & Gupta, M. (2008). Energy and exergy efficiency comparison of community-size and domestic-size paraboloidal solar cooker performance. *Energy for Sustainable Development*, *12*(3), 60–64. DOI: 10.1016/S0973-0826(08)60440-8

Khalifa, Y., & Al, E. (2025). The role of electric cooking in providing sustainable school meals in low-income and lower-middle-income countries. *The Lancet. Planetary Health*. Advance online publication. DOI: 10.1016/S2542-5196(25)00004-X PMID: 40054485

Kinyua, V., Wambugu, A., & da Silva, I. (2025). *D10.8 Report on field testing of standalone solar cooking appliance. European Union's Horizon 2020 Research and Innovation Program under Grant Agreement 963530., Long-Term Joint EU-AU Research and Innovation Partnership on Renewable Energy LEAP-RE. WP10 Task 10.2*. Strathmore University., Retrieved from https://www.leap-re.eu/wp-content/uploads/2025/02/LEAP-RE-D10.8-Report-on-field-testing-of-standalone-solar-cooking-appliance-V2.pdf

Kundapur, A. (2018). A treatise on Solar Cookers. [Kalashree, Hayagreva Nagara: International Alternate energy Trust.]. *Udupi*, *576*, 102.

Kundapur, A., & Sudhir, C. V. (2009). Proposal for a new world standard for testing solar cookers. *Journal of Engineering Science and Technology*, *4*(3), 272–281.

Lahkar, P., & Samdarshi, S. (2010). A review of the thermal performance parameters of box type solar cookers and identification of their correlations. *Renewable & Sustainable Energy Reviews*, *14*(6), 1615–1621. DOI: 10.1016/j.rser.2010.02.009

Lecuona, A., Nogueira, J. I., Ventas, R., Rodríguez-Hidalgo, M. C., & Legrand, M. (2003). Solar cooker of the portable parabolic type incorporating heat storage based on PCM. *Applied Energy*, *111*, 1136–1146. DOI: 10.1016/j.apenergy.2013.01.083

Lecuona-Neumann, A. (2017). *Cocinas solares. Fundamentos y aplicaciones. Herramientas de lucha contra la pobreza energética*. Editorial Marcombo.

Lecuona-Neumann, A., Nogueira, J., & Legrand, M. (2018). Photovoltaic Cooking, Chapter 13. In I. Yahyaoui, *Advances in Renewable Energies and Power Technologies. Volume 1: Solar and Wind Energies*. Elsevier. DOI: 10.1016/C2016-0-04518-7

Lecuona-Neumann, A., Nogueira-Goriba, J., Famiglietti, A., & Rodríguez-Hidalgo, M., & Boubour. (2024). Photovoltaic Cooker with No Electronics or Battery. *Energies*, *17*, 1192. DOI: 10.3390/en17051192

Lugolole, R., Mawire, A., Lentswe, K., Okello, D., & Nyeinga, K. (2018). Thermal performance comparison of three sensible heat thermal energy storage systems during charging cycles. *Sustainable Energy Technologies and Assessments*, 37–51. DOI: 10.1016/j.seta.2018.09.002

Mawire, A., & Abedigamba, O. (2025). Experimental comparison of erythritol and erythritol-granite pebble mixtures as heat storage materials for solar cooking. Green Technology & Innovation. *Green Technology & Innovation*, *1*(1), 6191. DOI: 10.36922/gti.6191

Mullick, S. C., Kandpal, T. C., & Saxsena, A. K. (1987). Thermal test procedure for box type solar cookers. *Solar Energy*, *39*, 353–360.

Nahar, N. (2003). Performance and testing of a hot box storage solar cooker. *Energy Conversion and Management, 44*, 1323–1331. DOI: 10.1016/S0196-8904(02)00113-9

Odoi-Yorke, F., Baah, B., & Opoku, R. (2025). Scientific mapping of experimental research on solar cookers: Global trends, evolution, and future directions. *Solar Energy Advances, 5*, 100093. DOI: 10.1016/j.seja.2025.100093

Ossei-Bremanga, R. N., & Al., E. (2024). Choice modelling of eCooking adoption by households in Ghana. *International Journal of Sustainable Engineering, 17*(1), 1–13. DOI: 10.1080/19397038.2024.2327832

Rahul Kashyap, S., Pramanik, S., & Ravikrishna, R. (2023). A review of solar, electric and hybrid cookstoves. *Renewable & Sustainable Energy Reviews, 188*, 113787. DOI: 10.1016/j.rser.2023.113787

Rohsenow, W., Harnett, J., & Cho, Y. (1998). *Handbook of Heat Transfer 3 rd.* McGraw-Hill.

Sarangi, A., Sarangi, A., Sahoo, S., Nayak, J., & Mallik, R. (2024). Advancements and global perspectives in solar cooking technology: A comprehensive review. *Energy Nexus, 13*, 100266. DOI: 10.1016/j.nexus.2023.100266

Sillé, R. Z., & van Duijsen, P. (2023). Low-cost Converter Design to Interface Solar to Inductive Cooking in Rural Areas. *International Conference on Electrical, Computer and Energy Technologies (ICECET).* Cape Town-South Africa: IEEE. DOI: 10.1109/ICECET58911.2023.10389500

Vipin, Singh, N., & Yadav, A. (2025). Fabrication and experimental investigation of a novel sensible heat storage system for indoor solar cooking. *Journal of Energy Storage, 117*, 116201. DOI: 10.1016/j.est.2025.116201

Worls Health Organization (WHO). (2022, January 20). *WHO publishes new global data on the use of clean and polluting fuels for cooking by fuel type*. Retrieved March 3, 2025, from WHO: https://www.who.int/news/item/20-01-2022-who-publishes-new-global-data-on-the-use-of-clean-and-polluting-fuels-for-cooking-by-fuel-type

Yayhyaoui, E. (2018). *Poor People's Energy Outlook 2018: Achieving Inclusive Energy Access at Scale*. Practical Action Publishing., DOI: 10.3362/9781780447544

ENDNOTES

1 Broken parenthesis ⟨⟩ represent functional dependence.

2 Variables are represented with italics characters and unit with normal characters.

APPENDIX A

A_a: Aperture area [m^2].
A_{ref}: Reference area for heat losses [m^2].
C: Heat capacity [J K^{-1}].
E: Energy [J].
G_T: Global solar irradiance on the plane of the solar panel [W m^{-2}].
h: Heat transfer coefficient [W m^{-2} K^{-1}].
k: Heat conductivity [W m^{-1} K^{-1}].
L: Length [m]
L_w: Water latent heat of vaporization [J kg^{-1}]
$\dot{m}$: Mass flow rate [kg s^{-1}]
R: Electrical resistance [Ω]
T: Temperature [ ]
t_B: Time to boil water [hr]
t_{coo}: Characteristic time for cooling [hr]
U: Global heat transfer coefficient [W m^{-2} K^{-1}].
W: Power; either solar W_s, delivered by the PV solar panel W_{PV}, or thermal W_{Th} [W].
δ: Thickness [m]
η: Energy efficiency [%]
σ: Root-mean square deviation from statistical hypothesis [%].
θ: Overtemperature [ ]

APPENDIX B

Figure 1. (a) Sample of the three commercial PTCs used. (b) A typical bare PTC element. (c) Diagram of the utensil and pot. (d) Electrical circuit for the main switch Sw0 and the three PTC switches.

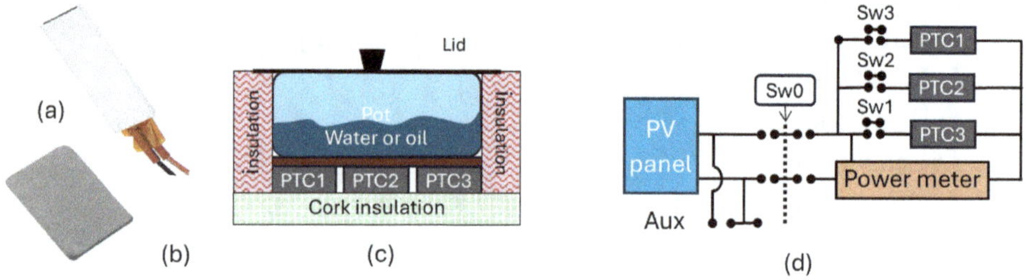

Figure 2. General view of the testing platform in the shadow of the solar panel showing the Pico technology® TC-08. (a) The bare pot on the mPVSC. (b) The same with the insulating lateral jacket.

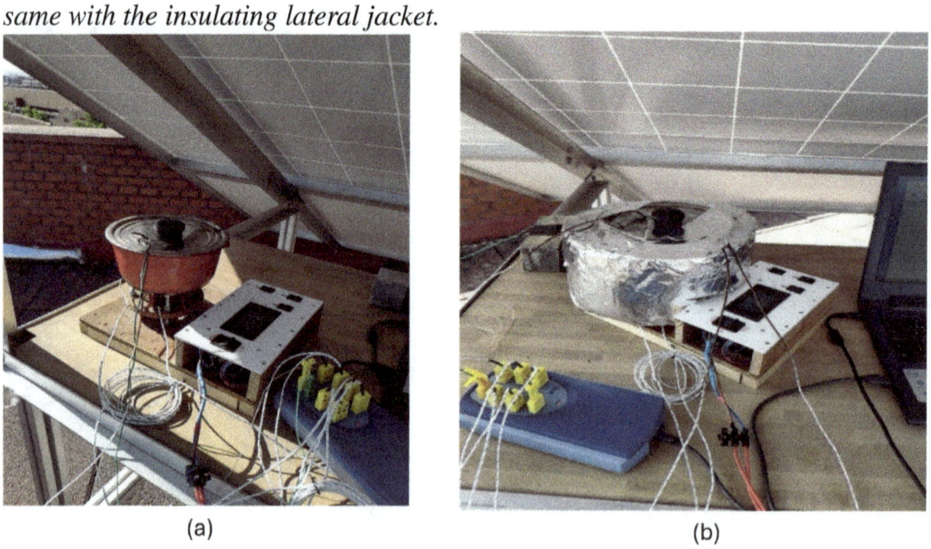

Figure 3. Utensil with the towels' insulation of the pot and hot plate, in place.

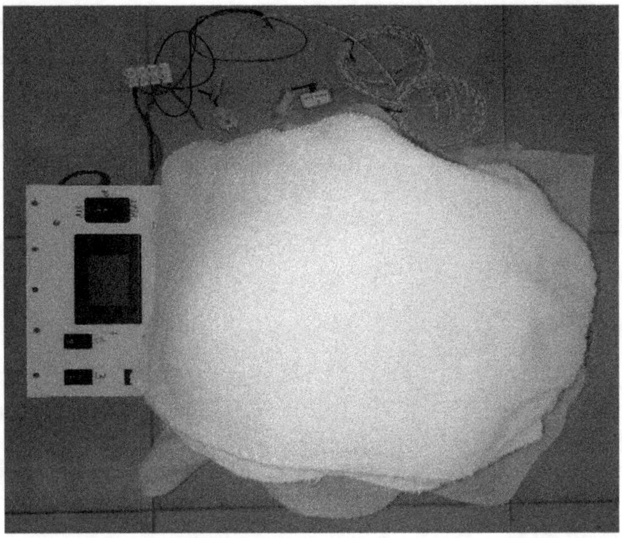

Figure 4. Evolution over time of the uninsulated (bare) pot on the mPVSC utensil over time during heating and successive cooling. Numbers, color coded, at the end of curves indicate end values. In the middle, the labels indicate maximum values. The – numbers – indicate the number of PTCs connected after this moment. EoC indicates the end of cooking, 60 °C.

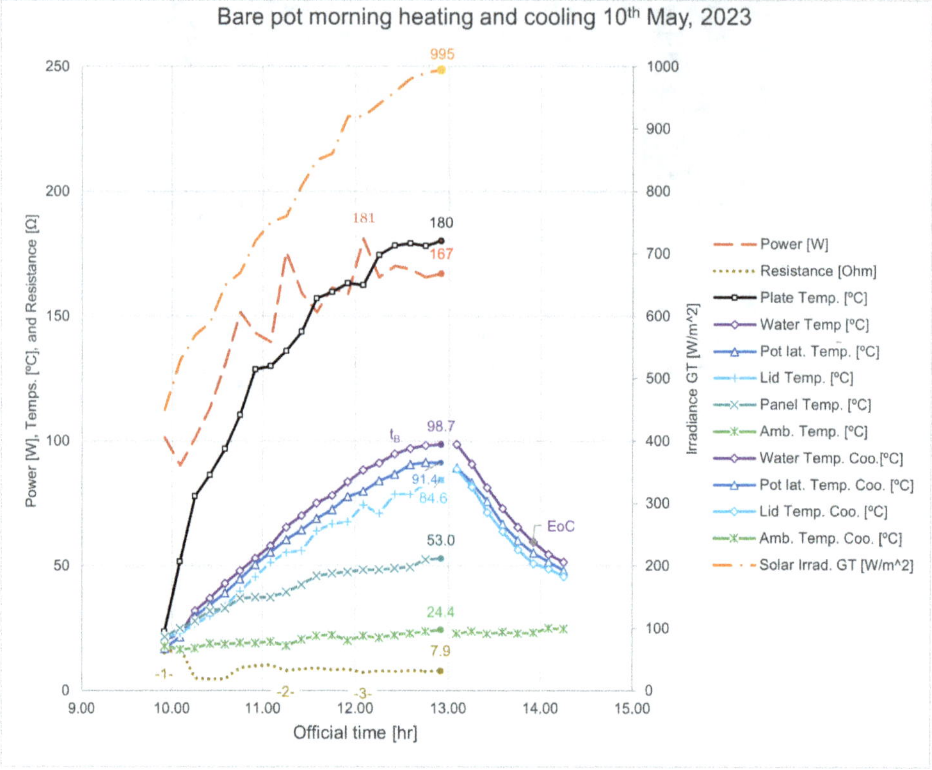

Figure 5. Evolution over time of the jacket insulated mPVSC utensil during reheating and successive cooling on May 10th. Numbers, color coded, at the end of curves indicate end values. In the middle, indicate maximum values. The – number – indicates the amount of PTCs connected after this moment. EoC indicates the end of cooking, 60 °C.

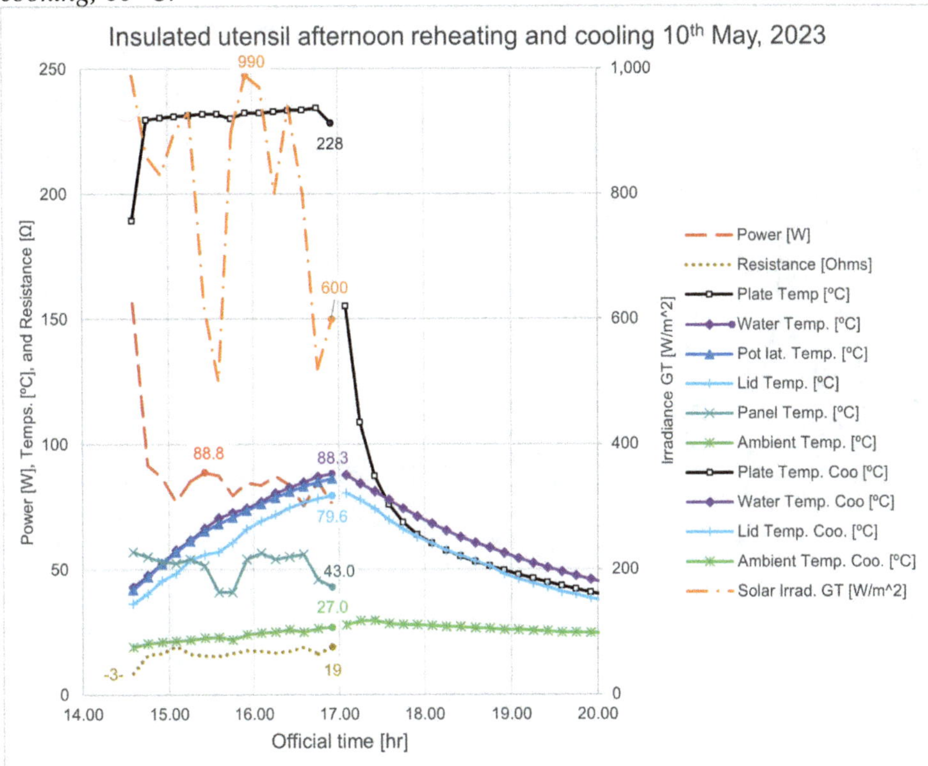

Figure 6. Evolution over time of the jacket-insulated mPVSC utensil during continuous heating and successive cooling on May 25th. Numbers, color coded, at the end of curves indicate end values. In the middle, indicate maximum values. The – numbers – indicates the amount of PTCs connected after this moment. EoC indicates the end of cooking, 60 °C.

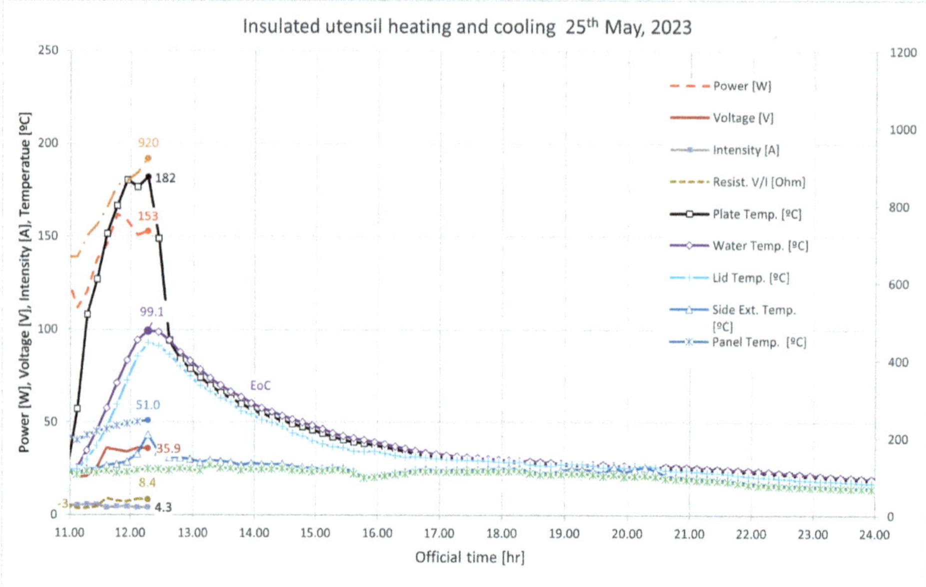

Figure 7. Heating and cooling of the heavily insulated pot with towels containing oil on May 9th. Numbers, color coded, at the end of curves indicate end values. In the middle, indicate maximum values. The – numbers – indicates the amount of PTCs connected after this moment. EoC indicates the end of cooking, 60 °C.

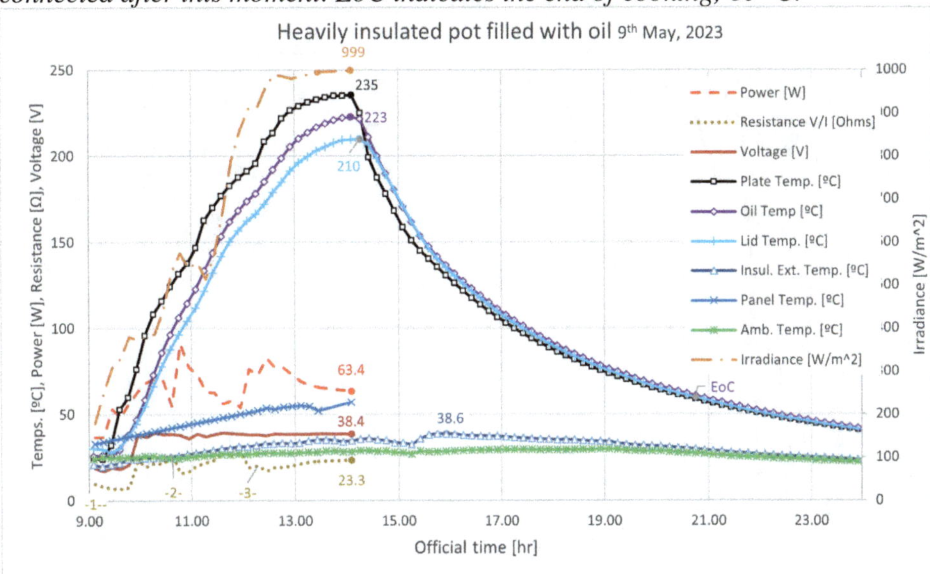

Figure 8. Evolution over time of the resulting voltage as the operator maximized power using the power meter during the heating and reheating processes on the three days, 10th and 25th with water, and the 9th with oil.

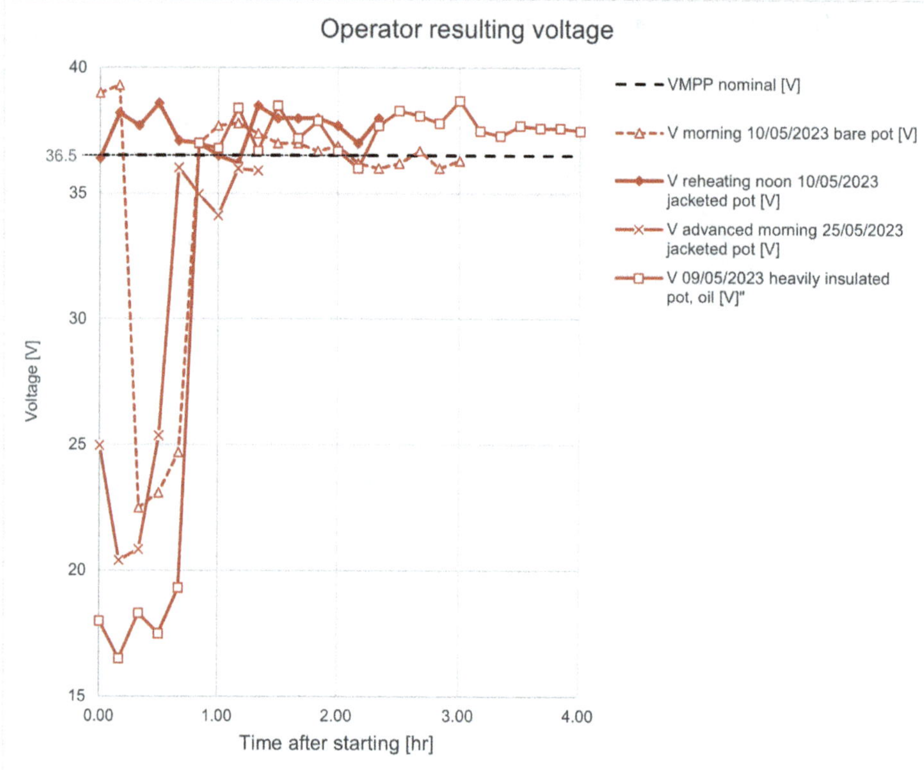

Chapter 2
Performance Evaluation of an Inclined Solar Box Cooker Under Different Climatic Conditions of Agadir City:
Design, Simulation, and Experimental Testing

Rachida Oaddi
https://orcid.org/0000-0002-9914-6143
Ibn Zohr University, Morocco

Mohamed Boulaid
Ibn Zohr University, Morocco

Rachid Safoui
Ibn Zohr University, Morocco

Rachid Tiskatine
Abdelmalek Essaadi University, Morocco

Abdellah Tihane
Ibn Zohr University, Morocco

Ahmed Aharoune
Ibn Zohr University, Morocco

Ahmed Ihlal
Ibn Zohr University, Morocco

ABSTRACT

Solar energy is increasingly vital in global strategies to reduce greenhouse gas emissions. Solar cooking is a key application, especially in rural areas where it lowers reliance on wood, improving health and reducing deforestation. This study focuses on designing, simulating, building, and testing a solar box cooker. Thermal

DOI: 10.4018/979-8-3373-2535-4.ch002

performance was simulated using TRNSYS, followed by outdoor tests in Agadir, Morocco, under varying conditions—different water loads, with/without reflectors, and with/without thermal load. A solar tracking system was implemented to keep the cooker aligned with the sun, improving efficiency. The prototype achieved a strong first figure of merit F1 (0.124 m^2°C/W), indicating high optical efficiency, and a second figure F2 (0.436), showing effective heat transfer. The cooker reached energy and exergy efficiencies of 19.36% and 10.12%, respectively. Overall, the tracking system significantly enhanced the performance of the solar box cooker.

1. INTRODUCTION

Solar energy is one of the most promising alternative to ensure the continuous growing demand of electricity and to meet basic human needs, while mitigating the effects of climate change by the reduction of greenhouse gases (GHG) emissions (Cuce & Cuce, 2013; Coccia et al., 2017). So far, several devices and applications have been developed to convert solar energy into other forms of useable energy, such as heat and electricity (Ghosh et al., 2017). Solar cooking is one of the most popular applications. There are many advantages in using solar cookers. One of them is that can take over of exploiting wood as a primary fuel for cooking especially in rural zones (Jebaraj & Srinivasa, 2015). The use of wood for cooking leads to significant ecological and health issues, including deforestation, burns, eye disorders, and lung diseases (Toonen, 2009). According to the World Health Organization (WHO), indoor air pollution from activities like cooking with firewood results in 1.6 million deaths annually (Rehfuess, 2006).

There are three main categories of solar cookers namely solar parabolic cookers, solar panel cookers and solar box cookers. Solar panel cooker is not highly appreciated since it provides a limited cooking power. Solar parabolic cooker presents risk of burning food if left without supervision for a period of time because of its high concentrated power and it needs continued adjustment to track closely the sun to reflect the incident solar radiation on the focus. Solar box cookers are the most popular ones, due to their simple design operation and lower costs. However, they take too much time to cook food in comparison with other types.

Numerous studies have focused on investigating and analyzing key factors affecting the solar box cooker. These factors include reflectors, absorbed plates, thermal insulation materials, glazing systems, cooking vessels, and thermal storage materials. The objective is to improve the effectiveness and efficiency of the solar box cooker by gaining a thorough understanding of key parameters. In a study conducted by Gianluca Coccia et al. (2017), a high concentration ratio solar box cooker with multiple reflectors was designed, manufactured, and tested (Coccia et al., 2017).

Experimental trials, including scenarios with and without loads such as water and peanut oil, were conducted to assess the maximum cooker temperature. The findings indicated that the cooker demonstrated the capability to achieve high temperatures with commendable optical efficiency and thermal insulation. Additionally, Lameck Nkhanjera et al. (2017) conducted a review focusing on heat storage materials and the cooking performance of solar cookers equipped with heat storage (Nkhonjera et al., 2017). This research contributes valuable insights into the selection of materials for heat storage and their impact on the overall cooking performance of solar cookers. Soria-Verdugo (2015) assessed the effectiveness of a box-type solar cooker (Soria-Verdugo, 2015). In their research, they introduced a convective coefficient model for heat transfer, validated the model with experimental data, and used it to simulate the solar cooker's performance in various countries worldwide. Hilario Terres et al. (2014) proposed a mathematical model to determine the thermal function on a solar box cooker with internal reflectors (Terres et al., 2014). An evaluation of exergy regarding the heat conduction in glass covers of a solar box cooker was proposed by Terres et al. (2017). They found that a considerable amount of entering energy to solar cooker during the heating process is not used (Terres et al., 2017). A thermodynamic review and a glimpse of diverse contributions in cooking technology through the solar box cooker have been presented by Saxena et al. (2011).

Harmim et al. (2010) conducted an experiment comparing the performance of two types of box-type solar cookers—one featuring a finned absorber plate and the other with a plain absorber plate (Harmim et al., 2010). Meanwhile, Hernández-Luna and Huelsz (2008) designed a solar oven specifically tailored for intertropical zones, focusing on the cooking process (Hernández-Luna & Huelsz, 2008). Abdulla et al. (2001) investigated the impact of solar cooker orientation on performance through experimental exploration, considering factors like the elevation angle of the sun, solar surface azimuth angle, and reflector tilt angle (Algifri & Al-Towaie, 2001). Nahar et al. (1994) conducted studies on a hot box solar cooker incorporating Transparent Insulation Materials (TIM) (Nahar et al., 1994). Their observations highlighted significantly improved performance compared to cookers lacking TIM. Going back, Mishra and Sabberwal Prakash (1984) evaluated the thermal performance of a solar cooker using readily available insulation materials in rural areas (Mishra & Sabberwal Prakash, 1984). Their goal was to minimize the cooker's cost, aiming for widespread adoption in rural Indian regions. Our contribution centers on the design, construction, and performance testing of a solar cooker adapted to the Agadir region. This area is unique as the sole region worldwide where Argan trees thrive. The escalating pressure on these trees has led to severe degradation of the forest ecosystems, exacerbating the desertification process in southern Morocco.

2. THERMAL PERFORMANCE ANALYSIS AND TEST PROCEDURES OF SOLAR COOKERS

2.1. Cooking Power

To assess the effectiveness of solar box cookers, various globally accepted parameters have been proposed and utilized. These include standard cooking power (P_s), first figure of merit (F_1), second figure of merit (F_2), and cooker thermal efficiency (η). Funk (2000) (Funk, 2000) introduced a testing method to estimate cooking power, outlined as follows:

$$P = \frac{(mc)\Delta T}{\Delta t} \quad (1)$$

Where (mc), represents the product of the fluid's mass and its specific heat capacity, ΔT denotes the temperature difference of the fluid, and Δt corresponds to the time interval set at 600 seconds. The author introduced a comprehensive metric termed "standard cooking power," defined by Equation (2) in their study.

$$P_s = P\frac{G_{ref}}{G_{av}} \quad (2)$$

Where G_{av} is the average solar irradiance for each time interval and G_{ref} is the reference illumination intensity level which is equal to 700 W/m^2. The cooking power equation appeared to be independent of location.

2.2. First and Second Figures of Merit

Mullick et al (Mullick et al., 1987) introduced a standardized testing protocol for box-type solar cookers. Within this procedure, two figures of merit are calculated. The first figure of merit F_1 is derived from a test conducted without any load. Experimentally, F_1 is determined using Equation (3).

$$F_1 = \frac{T_{ps} - T_{as}}{H_s} \quad (3)$$

Here, T_{ps} represents the stagnation temperature of the absorber plate, while T_{as} and H_s denote the ambient temperature and solar radiation on a horizontal surface, respectively, at the moment the stagnation temperature is attained.

Meanwhile, the second figure of merit F_2, is determined by heating a specified mass of water and is calculated using the following expression:

$$F_2 = \frac{F_1 (mc)_w}{A_p \Delta t} \ln \left[\frac{1 - \frac{1}{F_1}\left(\frac{T_{w1} - \bar{T}_a}{\bar{H}_s}\right)}{1 - \frac{1}{F_1}\left(\frac{T_{w2} - \bar{T}_a}{\bar{H}_s}\right)} \right] \tag{4}$$

Here, A_p represents the solar cooker's surface area, and Δt is the time interval (in seconds) required to elevate the water temperature from T_{w1} to T_{w2}. Additionally, $\bar{T}_a$ and $\bar{H}_s$ denote the average ambient temperature and horizontal solar radiation, respectively, observed over the time interval Δt.

2.3. Standard Boiling Time and Characteristic Curve of the Cooker

As per the guidelines established by Mullick et al (Mullick et al., 1987), the theoretical sensible heating time t is the anticipated time needed to heat the water within each 10-minute interval of the experiment. This can be calculated using the following expression:

$$t = -\frac{F_1 (mc)_w}{F_2 A_p} \ln \left[\frac{1 - \frac{(T_{w1} - \bar{T}_a)}{F_1 \bar{H}_s}}{1 - \frac{(T_{w2} - \bar{T}_a)}{F_1 \bar{H}_s}} \right] \tag{5}$$

The boiling time t_{boil} is the anticipated duration required to bring a specified volume of water to boil. This calculation is based on the recommended expression provided by Sethi et al (Sethi et al., 2014).

$$t_{boil} = -\frac{F_1 (mc)_w}{F_2 A_p} \ln \left[1 - \frac{(T_{boil} - \bar{T}_a)}{F_1 \bar{H}_s} \right] \tag{6}$$

3. ENERGY AND EXERGY CONCEPTS

Many scientists have been looking into how solar systems work using the rules of thermodynamics. This topic has caught the interest of many authors, and Bejan (Bejan, 1987) and Petela (Petela, 2003). Thermodynamic analysis is a useful way to get accurate and important information about how efficiently energy is used and the losses that happen due to irreversibility in a real situation.

3.1. Energy Efficiency

Energy analysis, relying on the First Law of Thermodynamics, primarily focuses on the amount of energy utilized and the efficiency of energy processes. This makes energy analysis well-suited for sizing and examining systems that operate using a singular form of energy.

To conduct an energy analysis of a solar cooker, it is necessary to assess the quantities of input and output energy. Assuming the negligible influence of kinetic and potential energy terms, and considering a steady-state flow process over a finite time interval, the overall energy balance equation for the solar cooker can be expressed as:

Energy input= Energy output+ Energy loss (7)

The energy input to the box solar cooker corresponds to the solar radiation energy per unit area of the cooker. This can be determined using the following calculation:

$$E_i = \overline{H}_s A_p \Delta t \tag{8}$$

The assessment of energy output from solar cookers focused exclusively on the sensible heat gained by the contents within the cooking vessel. The computation of solar cooker energy output relied on factors such as the variation in water temperature, mass of the water, and the specific heat capacity of water within the black painted vessel.

To compute the energy output, the variation in water temperature for each interval was multiplied by the product of the water's mass and its specific heat capacity. The formula for assessing the energy output of the solar cooker can be expressed as follows:

$$E_o = (mc)_w \left(T_{w2} - T_{w1}\right) \tag{9}$$

The energy efficiency is determined by the ratio of the energy output to the energy input of the solar cooker. This ratio is calculated using the following formula:

$$\eta = \frac{E_o}{E_i} = \frac{(mc)_w \left(T_{w2} - T_{w1}\right)}{\overline{H}_s A_p \Delta t} \tag{10}$$

3.2. Exergy Efficiency

Exergy analysis, primarily grounded in the Second Law of Thermodynamics, focuses on evaluating the quality of transferred energy. Engineers widely recognize exergy analysis as a powerful tool for appraising both the thermodynamic and economic performance of diverse systems (Petela, 2003). This method is valuable for improving system efficiency by quantifying energy waste and losses. In the scenario of a steady-state flow process over a finite time interval, the comprehensive exergy balance equation for the solar cooker can be expressed as follows the all-encompassing exergy balance equation for the solar cooker can be stated as follows:

Exergy input= Exergy output+ Exergy loss (11)

The widely accepted expression for calculating the exergy input to the solar cooker, as proposed by Petela (Petela, 2003), was utilized in the computation.

$$\Xi_i = \overline{H}_s \Delta t \left[1 - \left(\frac{4T_a}{3T_s}\right) + \left(\frac{T_a}{3T_s}\right)^4\right] A_p \tag{12}$$

The thermal exergy of heat energy at a given temperature (T) is then calculated based on this temperature.

The sun has a blackbody temperature of 5762 K, meaning it emits a concentrated spectrum of light primarily in the 0.3–3.0 μm wavelength range (KREITH et al., 2000). Although the sun's surface temperature (Ts) can vary on Earth due to this spectral distribution, a commonly used value for T_s is 5800 K.

The thermal exergy of heat energy at temperature T is then:

$$\varepsilon = \int_{T_0}^{T} m c_p \left(1 - \frac{T_0}{T}\right) dT \tag{13}$$

Where T_0 is the Reference temperature or dead state temperature (often the ambient temperature). This is the temperature of the surroundings with which the system eventually equilibrates.

Equation (9) can be applied for no isothermal processes. Thus, the thermal exergy content of water at temperature T_{w1} can be calculated by:

$$\varepsilon_w(T_{w1}) = (mc)_w \left[(T_{w1} - T_0) - T_0 \ln\left(\frac{T_{w1}}{T_0}\right)\right] \tag{14}$$

When the temperature of the water is increased to temperature T_{w2}, exergy increase is then:

$$\Delta \varepsilon_w = \varepsilon_w(T_{w2}) - \varepsilon_w(T_{w1}) \tag{15}$$

Thus, the exergy output from the solar cookers can be calculated using equation 16:

$$\Xi_o = (mc)_w\left[(T_{w2} - T_{w1}) - T_{ra}\ln\left(\frac{T_{w2}}{T_{w1}}\right)\right] \tag{16}$$

It is important to highlight that, in calculating the output exergy of the box solar cooker using equation (14), the value of T_{ra} is considered as the reference ambient temperature during the specified time interval.

Exergy efficiency serves as a metric comparing the useful energy output to the energy input required (Kotas, 1985). In the context of a solar cooker, exergy efficiency specifically denotes the ratio of the exergy gained by the solar cooker (exergy output) to the exergy of the incoming solar radiation (exergy input). This relationship is expressed mathematically by Equation 17:

$$\psi = \frac{\Xi_o}{\Xi_i} = \frac{(mc)_w\left[(T_{w2} - T_{w1}) - T_{ra}\ln\left(\frac{T_{w2}}{T_{w1}}\right)\right]}{H_s\Delta t\left[1 - \left(\frac{4T_a}{3T_s}\right) + \left(\frac{T_a}{3T_s}\right)^4\right]A_p} \tag{17}$$

4. SOLAR BOX COOKER MODEL

TRNSYS software is commonly used for solar box cooker modeling. Since solar cooker is not available in TRNSYS as a component, it is considered to be similar to a small building with a skylight window. The main assumption is that the solar box cooker is a building without any ventilation, infiltration, heating or cooling system. It means that only the external conditions (wind velocity and direction, outdoor temperature, humidity, and solar radiation) affect the thermal behavior of the solar cooker.

The TRNSYS model for a solar box cooker comprises various components with designated inputs and outputs. Within the TRNSYS simulation studio, the model assesses the average power absorbed by the solar cooker over a defined time period. Figure 1 illustrates the model, showcasing the TRNSYS components and their interconnections.

Figure 1. Solar box cooker simulation diagram

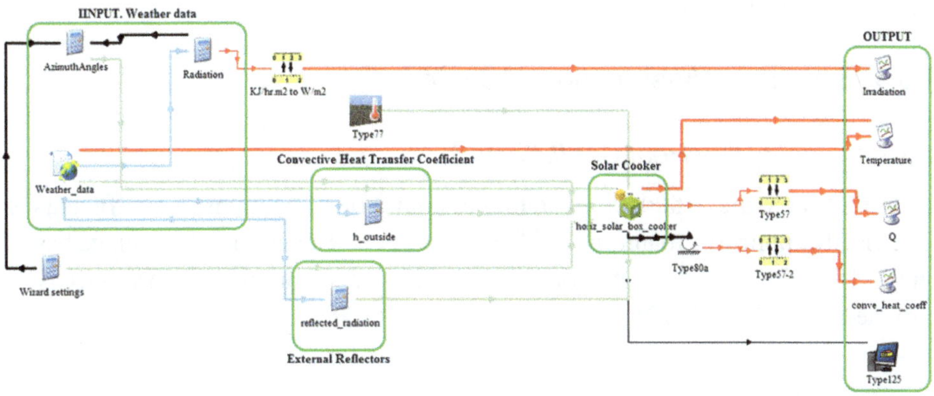

4.1. Weather Data

The TRNSYS model utilized the TRNSYS type 15-6 TM2 component for weather data modeling. This component functions by extracting data at consistent time intervals from an external weather data file (specifically, a meteonorm file was employed), interpolating the data at time steps of less than one hour, and making it available to other TRNSYS components.

4.2. Type 56

Type 56 refers to a building that has different areas with distinct temperature characteristics, known as thermal zones. The model considers information about the materials and thickness of walls and windows. Each air node has a homogenous temperature. Additionally, the model calculates the heat added to each air zone based on both direct and diffuse solar radiation, taking into account the properties of windows and heat transfer.

4.3. Type 65

Type 65 displays the selected variables at a specified time interval (ranging from one hour to one year), while the simulation progresses. This type is very widely used to visualize the graphic results of a simulation given the multitude of information that it is able to present simultaneously as well as its ability to detect possible malfunctions of the simulation and its configuration. If the program runs, a file containing the values of all selected variables will be created during the simulation.

The graphic data of this file can, subsequently, be transcribed into a "* .txt" or "* .xls" file for further processing.

4.3.1. Simulation Assumptions Adopted

4.3.1.1. Convective Heat Coefficients

In TRNSYS, the default computation of convection heat transfer coefficients is specifically applicable to internal surfaces. These coefficients are determined based on Eq (18), with the values of variables A and B corresponding to each surface type provided in Table 1 (1979).

$$h_{intern} = A\left(T_{surf} - T_{air}\right)^{B} \tag{18}$$

Table 1. Parameters for internal calculation of convective heat coefficient

Type of surface	A (kJ/h.m².K)	B (-)
Heated floor	7.2	0.31
Heated ceiling	3.88	0.31
Vertical wall	5.76	0.30

Moreover, since the equation Eq (18) cannot be used for the external surfaces of the building, the correlation presented by the Eq (19) has been adopted seeing that it takes into account the wind speed available in the meteorological file. (h_{extern}) is computed using the relation provided by Duffie and Beckman (DUFFIE & BECKMAN, 2013) (Equation 17):

$$h_{extern} = 5.6 + 3.8v \tag{19}$$

This equation is then linked to an input function created in Type 56.

4.4. Description of Cooking Prototype

4.4.1. Geometry

Considering the prevalent deprivation in various rural areas, we have developed an economically efficient solar furnace. The constructed solar cooker, depicted in Figure 2 and Figure 3, consists of two interconnected wooden boxes, each with a thickness of 15 mm. These boxes are nested within each other, leaving a 50 mm gap between them. The space between the boxes will be filled with insulating material to minimize heat loss. A 4 mm thick clear glass window is affixed to the top of the boxes within a wooden frame, facilitating heat retention through the greenhouse effect. The interior of the box is coated with black paint to optimize heat absorption.

Figure 2. Solar box cooker prototype.

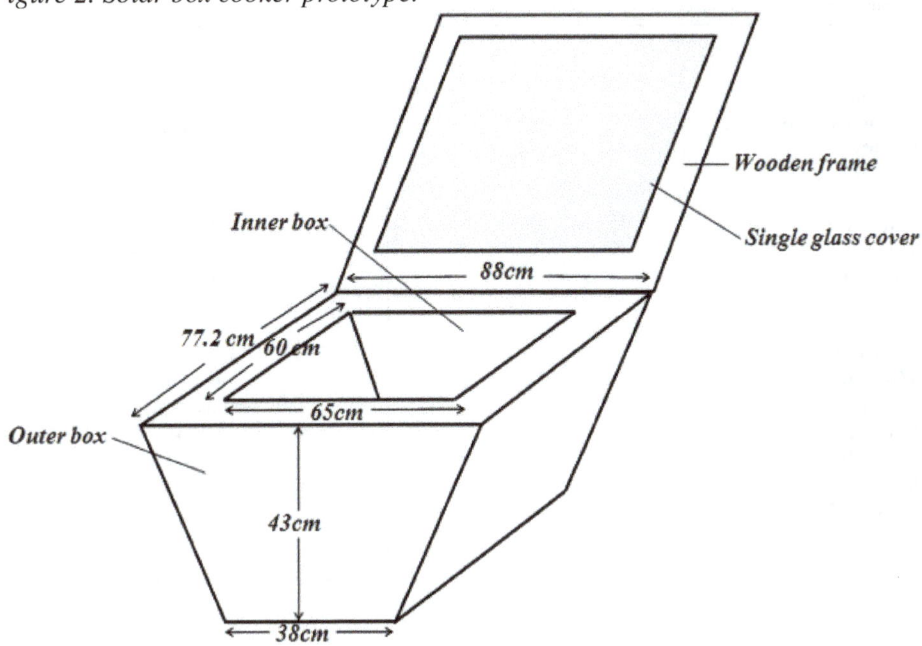

Figure 3. Solar box cooker prototype with Google SketchUp: (a) Horizontal surface; (b) Inclined surface.

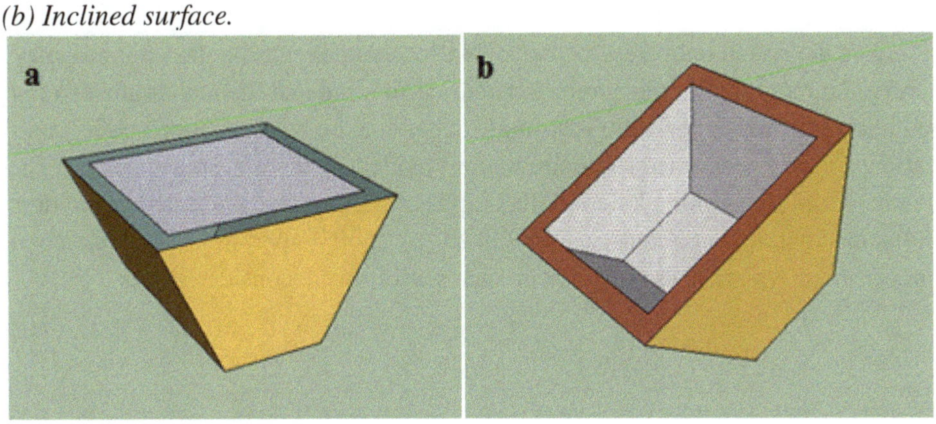

4.4.2. Materials Description

Table 2 gathers the main details of the designed solar cooker.

Table 2. Technical characteristics of the box cooker.

Parameters	Details
Casing	Wood
Material	0.015 m
Thickness	0.24 W/m.K
Thermal Conductivity	733.69 Kg/m^3
Density	929.13 J/Kg.K
Capacity	
Insulation Material	Polystyrene
Material	0.05 m
Thickness	0.03055 W/m.K
Thermal Conductivity	20 Kg/m^3
Density	1250 J/kg.K
Capacity	
Absorber plate	Aluminium
Material	0.004 m
Thickness	Matt black paint
Coating	0.93
Absorptivity	
Glaze	1
No. of glaze	Glass
Material	0.004 m
Thickness	

5. SIMULATION RESULTS

5.1. Incident Solar Radiation

Since the rise in temperature of any solar cooker depends on the solar radiation received through its opening zone and bearing in mind that the magnitude of solar radiation received by the latter is strongly depending on its orientation and its inclination, we have performed some simulations to identify the best configuration. The results of the simulation plotted in the graphs of Figure 4 are presented as incident radiation for horizontal and partly inclined (0°....60°) aperture area of the solar cooker. Four typical days (first day of each season) were considered.

Figure 4a. Hourly total incident solar radiation on each surface with time of the day; in winter (21 December)

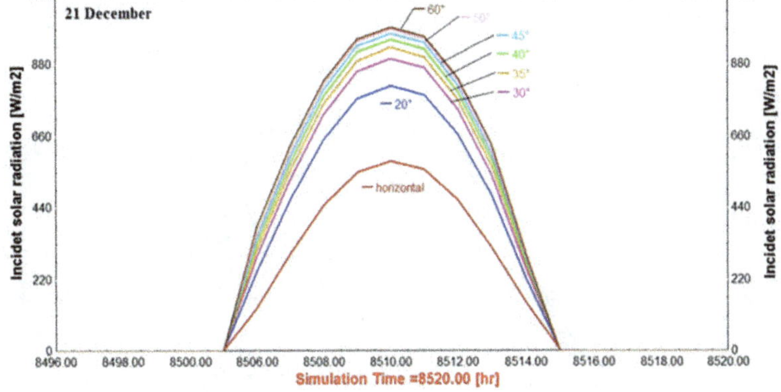

Figure 4b. Hourly total incident solar radiation on each surface with time of the day; in spring (21 March)

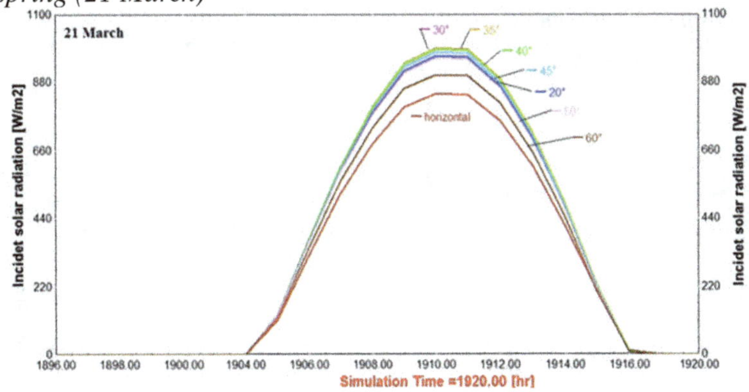

Figure 4c. Hourly total incident solar radiation on each surface with time of the day; in summer (21 June)

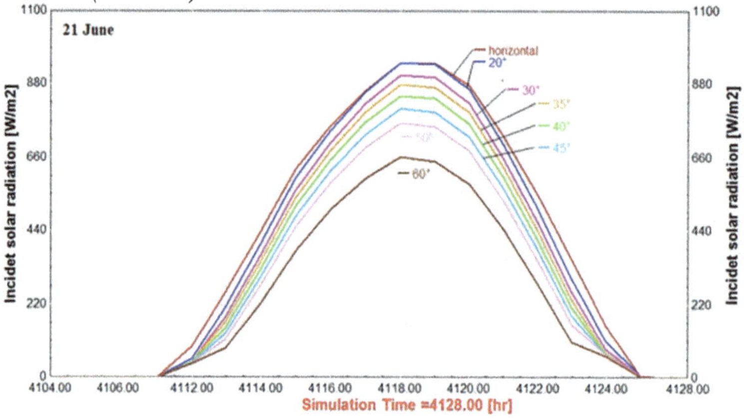

Figure 4d. Hourly total incident solar radiation on each surface with time of the day, in autumn (21 September)

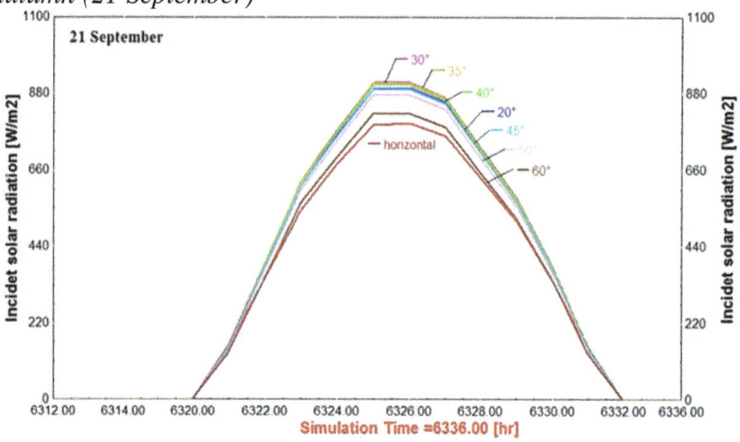

Figure 4 shows the hourly irradiation received by south facing surfaces of various slopes during equinoxes (21 March and 21 September) and solstices (21 June and 21 December). These values refer to latitude 30.4278° N. They provide a complete picture of the slope's influence on the solar irradiation received by a surface. A feature that this graphs display is that all the curves have a maximum corresponding to the solar noon.

During the winter solstice (21 December), we can notice that the curve representing the solar irradiation on horizontal surface have lowest values than that on inclined surfaces, especially surfaces with an inclination greater than 40°. This is because the sun is low in the horizon, so that the angle of incidence favors the inclined surface.

On the other hand, during the summer solstice (21 June), the horizontal surface receives the greatest amount of solar radiation and the most sloping surfaces receive the least amount, this is because the sun is much higher in the sky.

During the both equinoxes (21 March and 21 September), almost all inclined surfaces receive approximately the same amounts of solar radiation and the horizontal surface always receives the small amount.

As we can notice from the four graphs that the inclined surface with a slope of 30° always receives a significant amount of solar radiation. From the previous observations and based on the literature (Sethi et al., 2014; Heywood, 1971; MUJAHID, 1994), we can say that the optimum inclination angle for the aperture area of a solar box cooker corresponds to the latitude of the locality in question.

5.2. Temperature Profile

Based on the previous conclusion, the dimensions mentioned in Figure 2 and the properties of the materials illustrated in table 1, two configurations of the solar box cooker were simulated, the first with a horizontal surface (Figure 3.a) and the other with an inclined surface with a slope equal to 30 ° (Figure 3.b). The above mentioned four typical days were considered for the comparison. Figure 5 shows the temperature profile for the two configurations of the solar cooker and gives a general comparison between the temperatures inside the cooker and the temperatures of the absorber plates.

Figure 5a. Temperature profile in the case of a box cooker with horizontal and inclined aperture area (21 December)

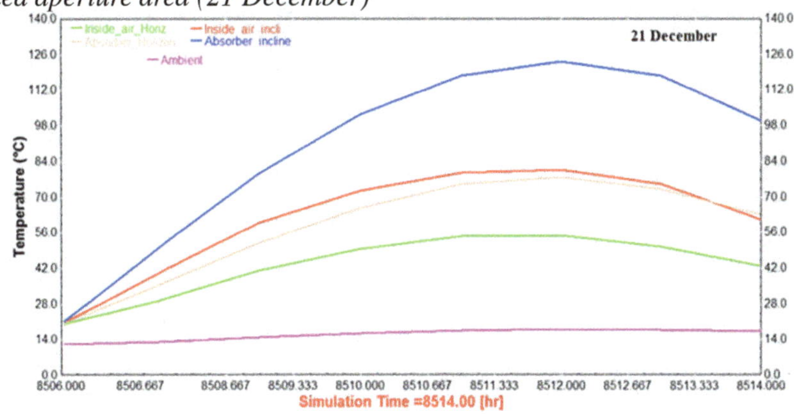

Figure 5b. Temperature profile in the case of a box cooker with horizontal and inclined aperture area (21 March)

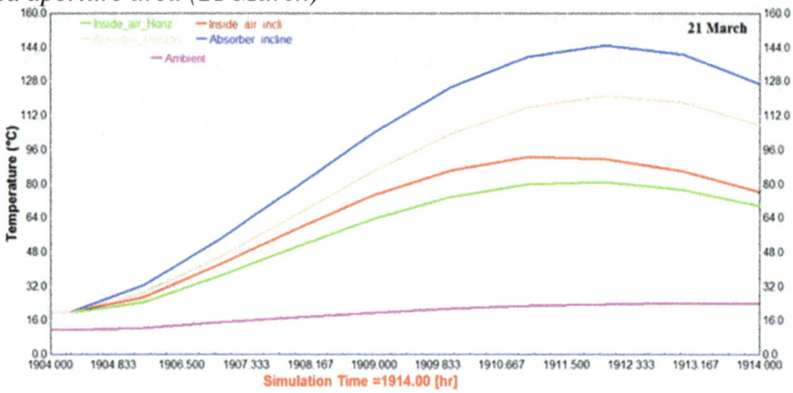

Figure 5c. Temperature profile in the case of a box cooker with horizontal and inclined aperture area (21 June)

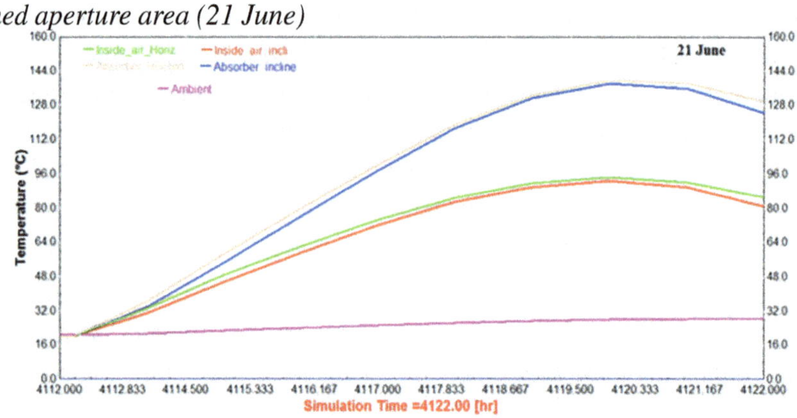

Figure 5d. Temperature profile in the case of a box cooker with horizontal and inclined aperture area (21 September)

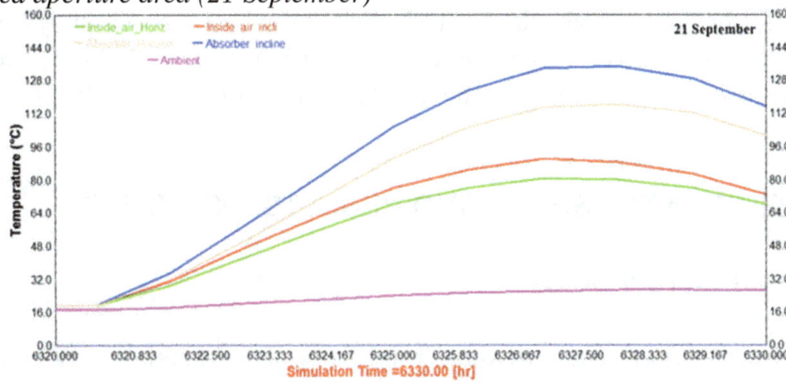

The solar radiation fallen on the aperture surface throughout the whole day act as a crucial heat source for the solar cooker. As it can be seen from Figure's 5 graphs, the inside and absorber plate temperatures are always high for the inclined solar box cooker compared to the cooker with a horizontal aperture area except during the summer solstice where both cookers reach almost the same temperatures.

Table 3 summarizes the temperature, solar irradiation and first figure of merit (calculated using equation 3) and values reached at the stagnation time by both solar cookers.

Table 3. Stagnation test values

	21 December	21 March	21 June	21 September
T_{ps_h} (°C)	78.06	121.5	140.1	117
T_{ia_h} (°C)	55	81.4	94.6	80.7
T_{as_h} (°C)	18.20	24.15	28.41	27.35
H_{s_h} (W/m²)	468.0	758	873.9	639.1
F_1 (m².K/W)	0.127	0.128	0.127	0.140
$T_{ps\_30°}$ (°C)	126.8	144.8	138.5	135.4
$T_{ia\_30°}$ (°C)	82.31	91.97	92.90	89.3
$T_{as\_30°}$ (°C)	18.20	24.15	28.39	27.27
$H_{s\_30°}$ (W/m²)	745.0	890.6	825.3	737.9
F_1 (m².K/W)	0.232	0.159	0.126	0.169

In Table 3, it is evident that both solar cookers possess first figure of merit (F_1) values exceeding 0.12. Our findings align with the international standard proposed by Mullick et al (Mullick et al., 1987), which stipulates that F_1 should fall within the range of 0.12 to 0.16. A higher F_1 value signifies superior optical efficiency and lower heat loss factor (Yettou et al., 2014). The inclined solar cooker consistently exhibits elevated F_1 values across the four observed days, attributed to its design featuring a south-inclined aperture area, resulting in enhanced solar radiation collection.

6. EXPERIMENTAL SETUP AND RESULTS DISCUSSION

6.1. Experimental Setup

Experimental tests were carried out on the rooftop of the Materials and Renewable Energies laboratory at the Faculty of Sciences, located at 30.4060 °N latitude and 9.5444 °W longitude. The tests ran from 10:00 a.m. to 5:00 p.m. Various measuring devices, as depicted in Figure 6, were employed during the experiments. Solar radiation intensity was quantified using a digital Pyranometer, with validation for values falling within the range of 450 W/m^2 to 1100 W/m^2.

Because our simulations confirmed that 30° collects significant solar energy, and this angle also matches Agadir's optimal latitude, we scientifically decided to build and test our solar box cooker with its opening inclined at 30° from the flat ground.

To measure temperatures at different points in the solar cooker, including the ambient air, interior air, absorber plate, and water, K-type thermocouples with an accuracy of ±0.5 °C were utilized. Acceptable conditions for the ambient temperature ranged between 20-35 °C, and wind velocity was required to be less than 1.0 m/s. All variables were systematically measured and recorded at 10-minute intervals.

The experimental tests were conducted under both stagnation and full load conditions, encompassing clear and semi-cloudy days. During the experimental tests, errors and measurement uncertainties are likely to occur, which may be caused by the type of instrument, climatic conditions, calibration methods, observation, and test methods.

Figure 6. Instruments used while testing (a) FI-109SM Solarimeter, (b) Data logger for measuring temperatures

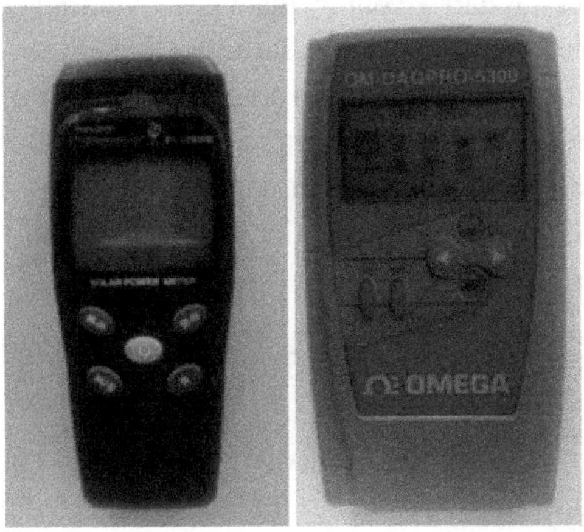

Figure 7. Image of the constructed solar box cooker (left), data logger with thermocouples during experimental test (right)

6.2. Determination of Performance Parameters

The thermal performance evaluation of the solar cooker involves two distinct tests: the stagnation test and the sensible heat test. The results from the stagnation test yield the initial figure of merit, denoted as (F_1). In this test, the temperature of the absorber plate is recorded along with the solar radiation on a horizontal surface and ambient temperature. The absorber plate's temperature initially rises and eventually reaches a stagnant state. At stagnation F_1 can be calculated using Equation (3).

The sensible heat test is carried out by placing a cylindrical cooking vessel filled with water and covered with a lid inside the solar cooker. According to the Indian Standard IS 13429, the aperture area must be 8 kg/m^2, evenly distributed among all cooking pots. Considering that the area of the cooker is 0.39 m^2, the total mass of required water is 3.12 kg in the present experiments. Results from the sensible heat test contribute to the calculation of the second figure of merit, denoted as F_2 using equation (4).

6.2.1. Temperature Profile for the Stagnation Test

The stagnation temperature results are associated with tests conducted without load and reflectors. In Figure 8, the typical variations in solar irradiation on a horizontal surface, ambient temperature, absorber plate temperature, and inside air temperature of the cooker are presented. These parameters were recorded during a stagnation test under cloudy sky conditions, characterized by intermittent cloud cover and no wind. Additionally, Figure 9 displays the test values for the implemented solar box cooker on a clear day.

Figure 8. Diurnal variation of solar irradiation, ambient temperature, measured plate temperature and inside air temperature of the cooker under stagnation test condition on cloudy day (15 September)

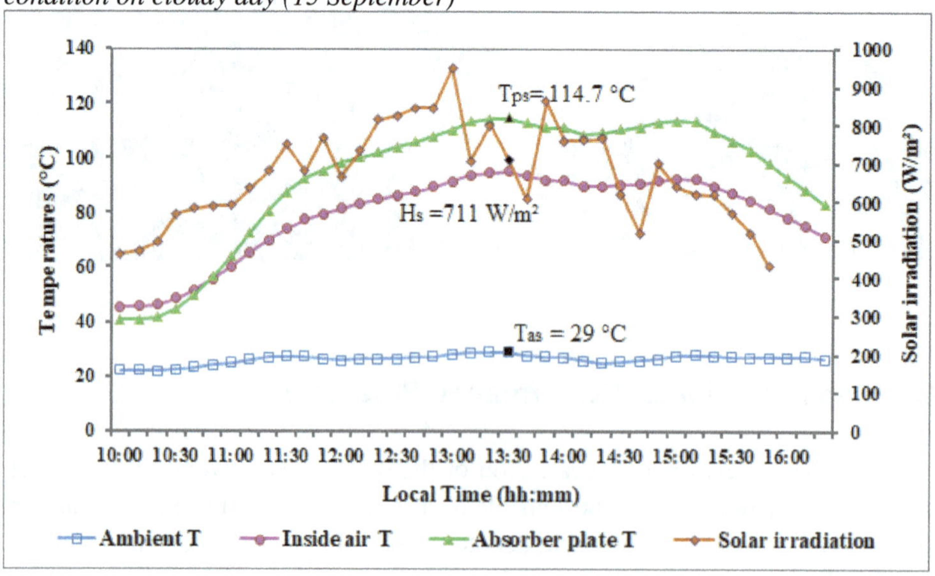

Figure 9. Diurnal variation of solar irradiation, ambient temperature, measured plate temperature and inside air temperature of the cooker under stagnation test condition on a clear day (26 September)

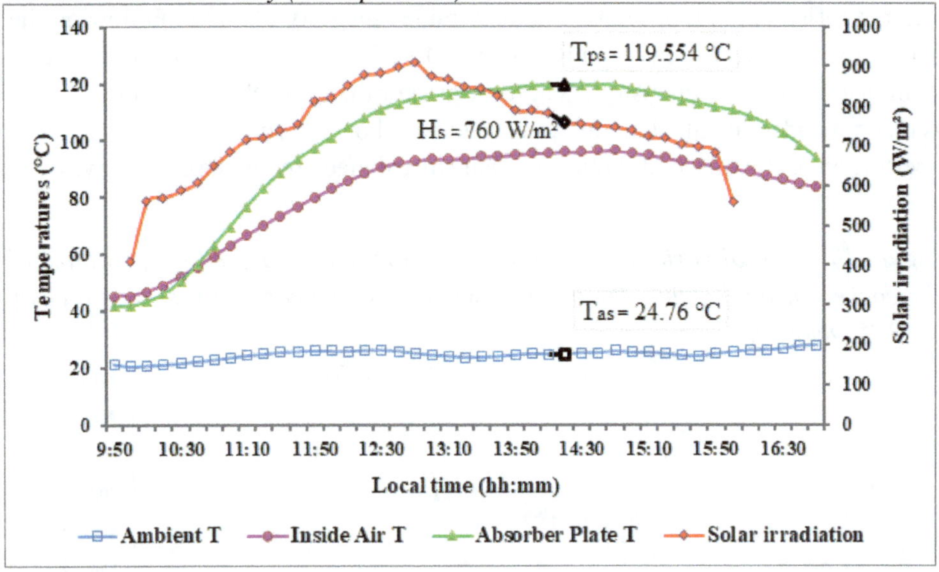

6.2.1.1. Determination of First Figure of Merit

Under cloudy sky conditions, the maximum absorber plate temperature and inside air temperature reached 114.7 °C and 95.13 °C, respectively, at 13:30 (Figure 8). These values were recorded at a solar irradiation level of 711 W/m^2 and an ambient temperature of 29.17 °C. The results obtained under cloudy sky conditions indicate the solar cooker's potential for cooking, as the inside air and absorber plate temperatures remain within acceptable ranges. Alternatively, it can be used for reheating previously cooked food.

Under clear sky conditions, the maximum absorber plate temperature was 119.554 °C, observed at 14:20 local time (Figure 9). This occurred at a solar irradiation of 760 W/m^2 and an ambient temperature of 24.76 °C. The first figure of merit F_1, was calculated to be 0.124 m^2 °C/W using equation (3). The thermal performance of the cooker was deemed satisfactory, earning it a grade-A classification. According to Indian standards, any designed solar box cooker with F_1 greater than 0.12 is categorized as grade A, while those with F_1 equal to or less than 0.12 are classified as grade B (Mirdha & Dhariwal, 2008).

6.2.2. Temperature Profile for the Sensible Heat Test

This test aims to find the second figure of merit, denoted as F_2. A solar cooker without reflectors is filled with a predetermined quantity of water. Following the standards set by the Bureau of Indian Standards (BIS), the experiment requires solar illumination with an intensity equal to or greater than 600 W/m^2. Additionally, it is recommended to calculate the water load based on 8 kg/m^2 of the solar cooker's aperture area, and this load should be evenly distributed among the cooking vessels.

Figure 10. Diurnal variation of solar irradiation and measured temperatures of the cooker during heating 3kg of water within a single cooking vessel on a clear day (05 October)

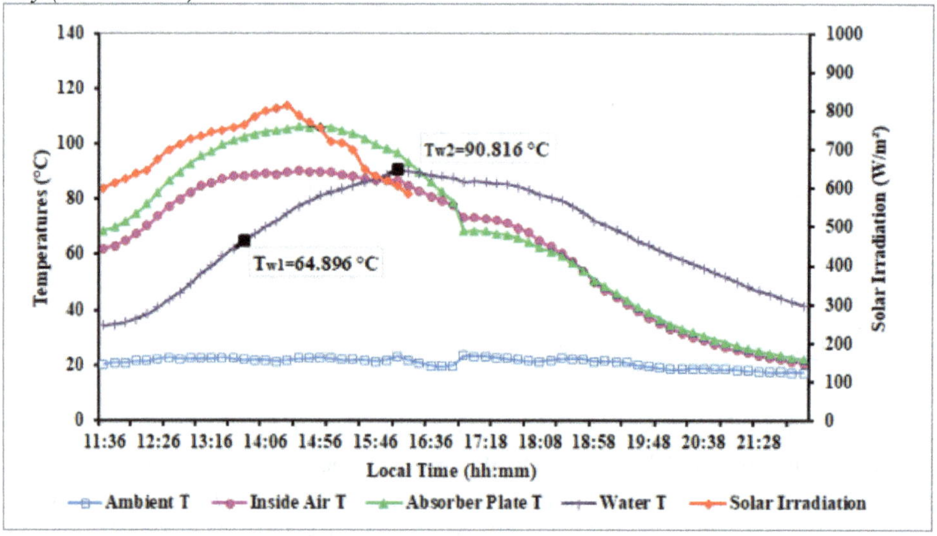

Figure 11. Diurnal variation of solar irradiation and measured temperatures of the cooker during heating 3kg of water within two cooking vessels on a clear day (10 October)

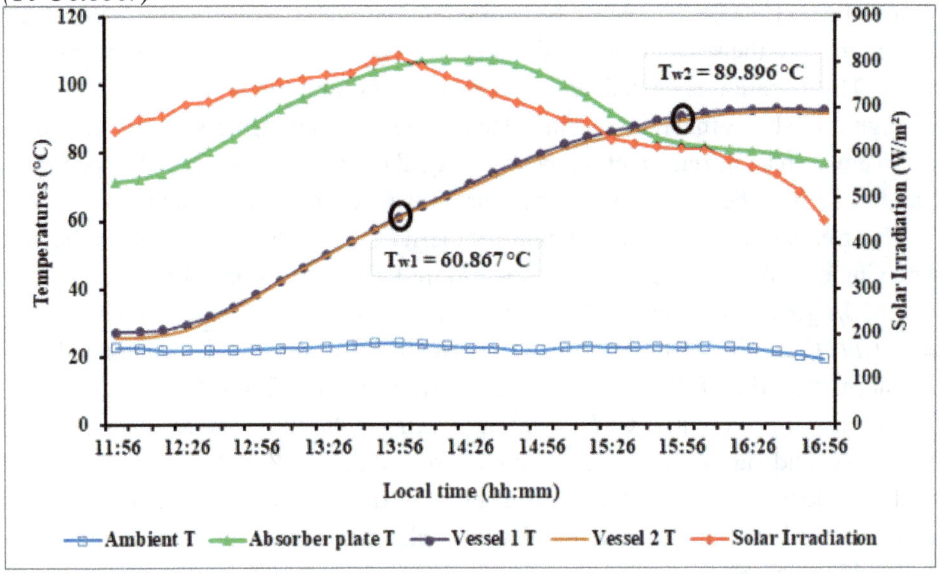

Figure 12. Standardized cooking power vs. temperature difference and the resulting regression line for the test using a single cooking vessel (left) and using two cooking vessels (right)

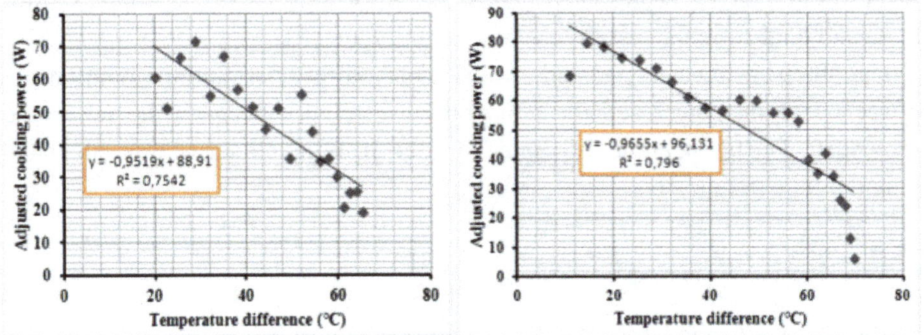

6.2.2.1. Determination of Second Figure of Merit

Figure 10 recorded data from an experiment illustrates the water temperature within a single cooking vessel, along with the inside air temperature, absorber temperature, ambient temperature, and solar radiation intensity during sensible heat tests. The weather conditions were predominantly clear and sunny. The measurements started at 11:30 am with the initial water temperature at 34.48 °C and the ambient temperature at 20.21 °C. After 140 minutes, the water temperature reached its maximum, and the solar cooker maintained the cooking temperature for over three hours. This duration is considered sufficient for cooking food.

Figure 11 shows that the experiment test using two cooking vessels was started at 11:56 am with an initial water temperature equal to 26 °C and ambient temperature equal to 22 °C. The maximum temperature was reached after 130 min.

Using equation (4), the second figure of merit F_2 is calculated as 0.452, for the test with single cooking vessel, using $F_1 = 0.124$ m^2 K/W with values of $m_w = 3$ kg, $C_{pw} = 4185$ J/kg.K, $A_p = 0.39$ m^2, $\Delta t = 140$ min, $T_{w1} = 64.896$ °C, $T_{w2} = 90.816$ °C, $\bar{T}_a = 22.677$ °C and $\bar{H}_s = 683.5$ W/m^2. The corresponding value of F_2 calculated from the thermal profile of the test using two cooking vessels (Figure 11) is found to be 0.436, using $T_{w1} = 60.867$ °C, $T_{w2} = 89.896$ °C, $\bar{T}_a = 23.29$ °C and $\bar{H}_s = 711$ W/m^2.

The second figure of merit, F_2, ranges from 0.254 to 0.490 depending on the load and the number of cooking vessels employed. A higher F_2 value signifies the efficiency of heat transfer from the absorber plate and inside air to the contents of the cooking vessels.

6.2.2.2. Determination of Cooking Power

The cooking power interval and standardized cooking power for the solar cookers were determined through the analysis of experimental data from the sensible heat tests, employing Equations (1) and (2). Subsequently, the standardized cooking power was graphically represented against the temperature difference for two cases: single and dual cooking vessels, as depicted in Figure 12.

The obtained linear regressions were employed to determine the relationship between cooking power and the temperature difference between the water and ambient air. The temperature difference for each 10-minute interval was calculated using the following expression.

$$T_d = T_w - T_a \tag{16}$$

The obtained standardized cooking power and linear regressions are $P_s = 88.91 - 0.9519T_d$ (left graph) and $P_s = 96.131 - 0.9655T_d$ (right graph) and the standards cooking power for a 50 °C temperature difference are 41.315 W and 47.856 W. The coefficients of determination (R^2) for the analysis are 0.7542 and 0.796. To ensure a consistent and uniform cooking power in the solar cooker during the cooking process, it is advisable for R^2 to exceed 0.75, as recommended by Funk (Funk, 2000). The obtained R^2 values meet this recommended standard.

6.2.2.3. Standard Boiling Time

Figure 13 displays the recorded experimental data for water temperature within two cooking vessels, inside air temperature, absorber temperature, ambient temperature, and solar radiation intensity during the sensible heat test.

Figure 13. Diurnal variation of solar irradiation and measured temperatures of the cooker during heating 3kg of water within two cooking vessels on a clear day (12 October)

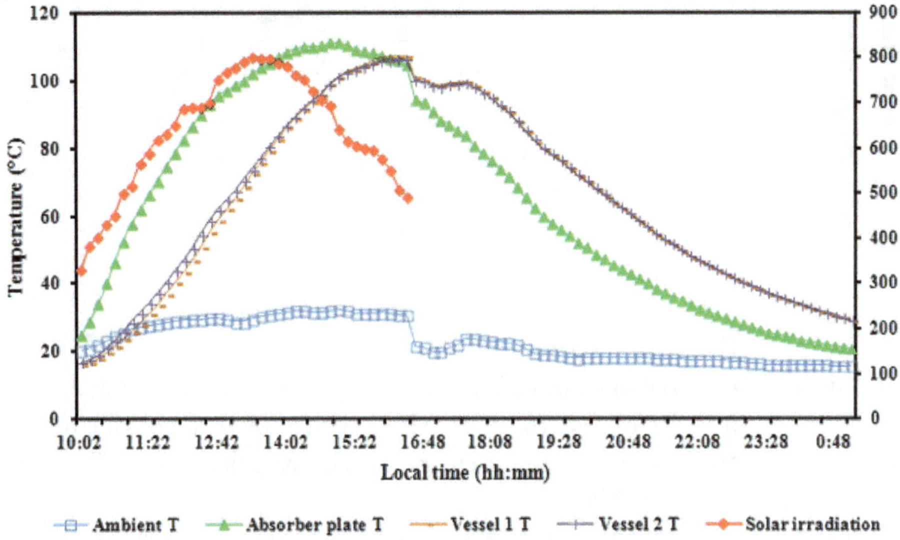

The weather conditions were mainly clear and sunny. The measurements commenced at 10:02 am with the initial water temperature at 16.6 °C and the ambient temperature at 20.35 °C. It took 300 minutes to reach the boiling temperature. The maximum water temperature of 106.7 °C was attained at 16:22, and the solar cooker maintained the cooking temperature for over 3 h 30 min.

The second figure of merit F_2 is calculated as 0.428, for this test, using F_1= 0.124 m^2 K/W with values of m_w = 3 kg, C_{pw} = 4185 J/kg.K, A_p = 0.39 m^2, Δt = 90 min, T_{w1} = 60.491 °C, T_{w2} = 89.044 °C, $\bar{T}_a$ = 31.102 °C and $\bar{H}_s$ = 759.5 W/m^2.

Figure 14. Standardized cooking power vs. temperature difference and the resulting regression line for the test using two cooking vessels (boiling test)

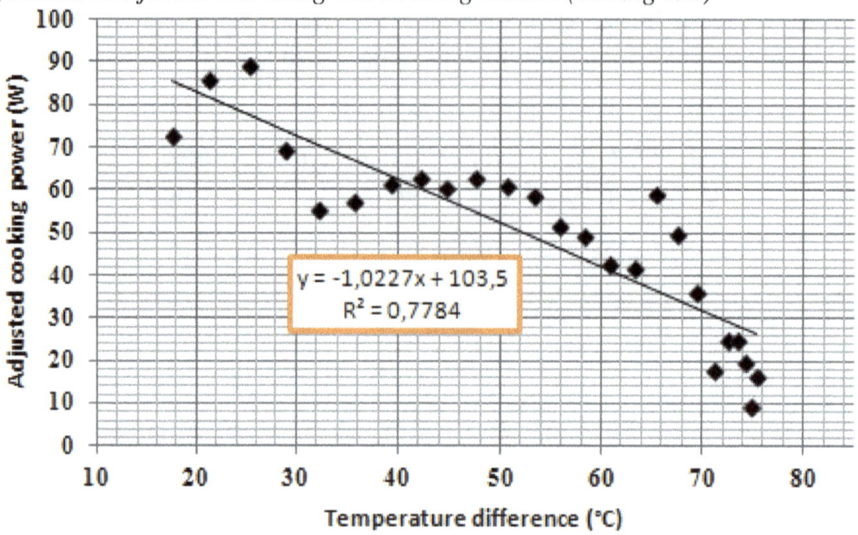

The standardized cooking power and linear regression obtained (shown in Figure 14) are represented by P_s= 103.5–1.0227T_d, with a standard cooking power of 52.36 W for a 50 °C temperature difference. The corresponding coefficient of determination R^2 value is 0.7784.

Experimentally, as depicted in Figure 13, it took 300 minutes for the water temperature to reach boiling temperature under ambient conditions of 32.415 °C and solar irradiation of 644 W/m². While, the calculated boiling time using equation (6), F_1 = 0.124 m² K/W and F_2= 0.428 was t_{boil} = 291 min, resulting in an absolute error of 9 minutes, This discrepancy is considered acceptable when compared to the 13-minute and 16-minute errors calculated by Sethi et al (Sethi et al., 2014) and Yettou et al (Yettou et al., 2023), respectively.

6.2.3. Energy and Exergy Efficiencies of the Solar Cooker

6.2.3.1. Energy and Exergy Output

Figure 15 illustrates the variation in the energy and exergy output with time of the studied solar cooker, calculated using equations (9) and (16) respectively.

Figure 15. Energy and Exergy outputs of the solar cooker with time

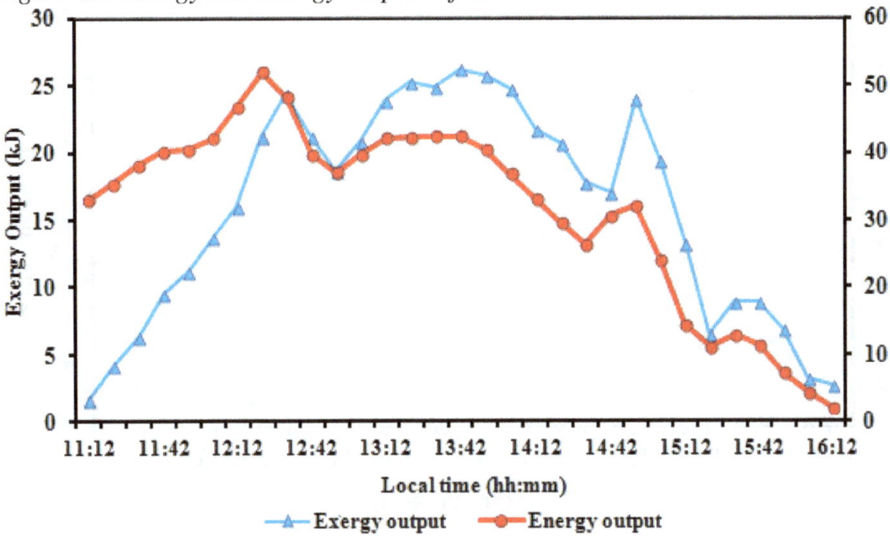

Throughout the study period, the energy output of the solar cooker ranged from 2.02 to 51.98 kJ, while the exergy output spanned from 1.68 to 26.21 kJ. Figure 15 illustrates that the energy output demonstrated a rapid increase in the initial hour, followed by a decline. On average, the daily energy output for the Solar Box Cooker (SBC) was 30.41 kJ.

A similar pattern was observed in the variation of exergy output, with an average daily exergy output for the SBC measuring 12.88 kJ. Notably, the study revealed that the energy output consistently surpassed the exergy output.

6.2.3.2. Energy and Exergy Efficiencies

Figure 16 shows the comparative results of energy and exergy efficiencies which were calculated using equations (10) and (17) respectively.

Figure 16. Energy and Exergy efficiencies of the solar cooker with time

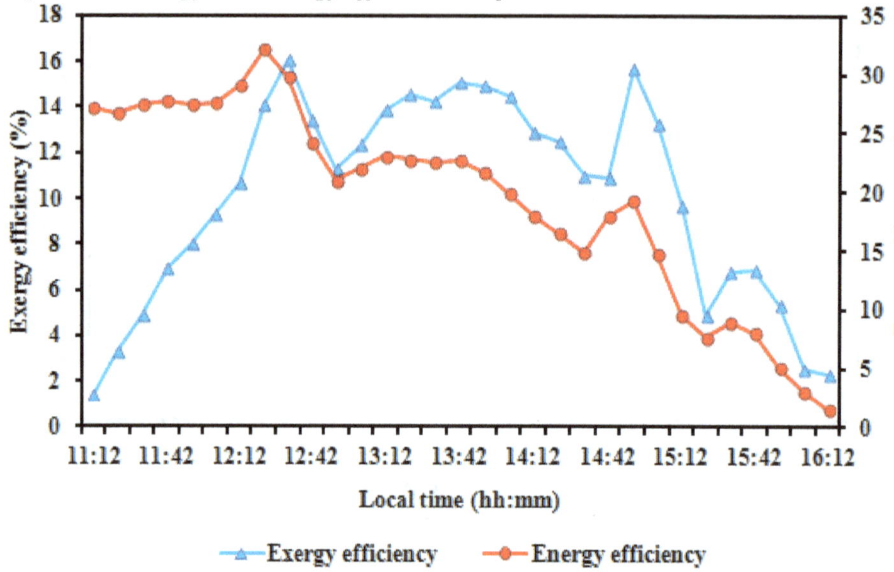

The solar cooker demonstrated energy efficiency ranging from 1.56% to 32.10%, with an average daily energy efficiency of 20.19%. These results are consistent with Todd and Miller's findings (1999), who reported average efficiencies of 8% for a basic cooker and 22% for the best-tested design. The outcomes indicate that the current solar cooker is well-designed and operates with efficient energy performance. On the other hand, exergy efficiency fluctuated between 1.41% and 15.66% during the same experimental period, with an average daily exergy efficiency of 10.12%. A comparison of energy efficiency with exergy efficiency provides valuable insights:

- The energy efficiency consistently surpassed the exergy efficiency, as anticipated. This is because the calculation of energy efficiency considers the total energy content of the hot water used as a heat storage fluid. In simpler terms, when calculating energy efficiency, the focus is on the quantity of energy transferred, while the quality of the energy transferred is not taken into account.
- The results shown above reveal that across all temperatures, the exergy efficiency of the solar cooker consistently remained lower than its energy efficiency.
- The calculated average daily exergy efficiency over the experimental period was found to be 10.12%. This result suggests that the solar box cooker investigated in this study demonstrates inefficiency in terms of exergy efficiency. This observation is consistent with the findings of Krane (Krane, 1987), who

also concluded that sensible heat energy storage systems tend to be inherently inefficient in terms of exergy efficiency.

6.2.4. Temperature Profile for the Stagnation Test With Tracking System

A Solar Tracker is a device designed to follow the sun's movement as it travels from east to west each day. These trackers are employed to ensure that solar collectors or panels consistently face the sun as it moves across the sky. This practice effectively enhances the incident solar irradiance, leading to improved energy output in the form of heat or electricity.

The objective of using a solar tracker in this context is to maintain the solar box cooker aperture area perpendicular to the sun throughout the day, optimizing its efficiency. The support platform on which the box is positioned is connected to a DC motor controlled by an Arduino board. This mechanism allows for precise adjustment to keep the solar box cooker aligned with the sun's position.

Figure 17. DC motor connected to the support on which the solar box cooker is placed

The test was carried out by taking the temperature variation of the absorber plate at three different points. Two days of testing with almost the same climatic conditions were taken into consideration. Figure 18 shows the temperature profile of the absorber plate during a test without using the tracking system. As can be seen, the three parts of the plate do not reach the same temperature, this is due to the fact that the East and West sides are shaded by the box's walls. On the other hand, Figure 19 presents the temperature profile of the absorber plate during a test using the tracking system. The entire surface of the plate reaches almost the same temperature; this demonstrates that the tracking system keeps the aperture area of the solar cooker perpendicular to the sun during the whole test.

Figure 18. Temperature profile in three different positions of the absorber plate without tracking system (15 October)

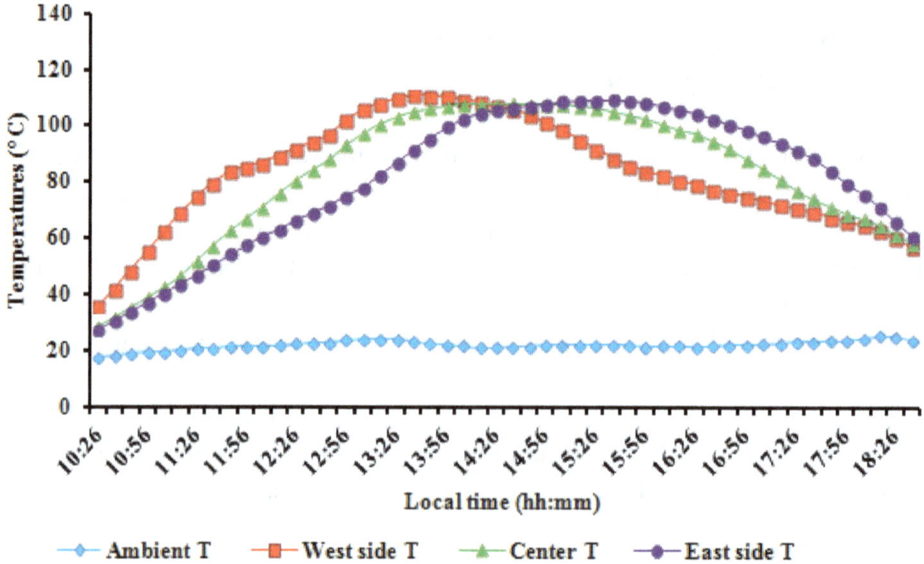

Figure 19. Temperature profile in three different positions of the absorber plate with tracking system (19 October)

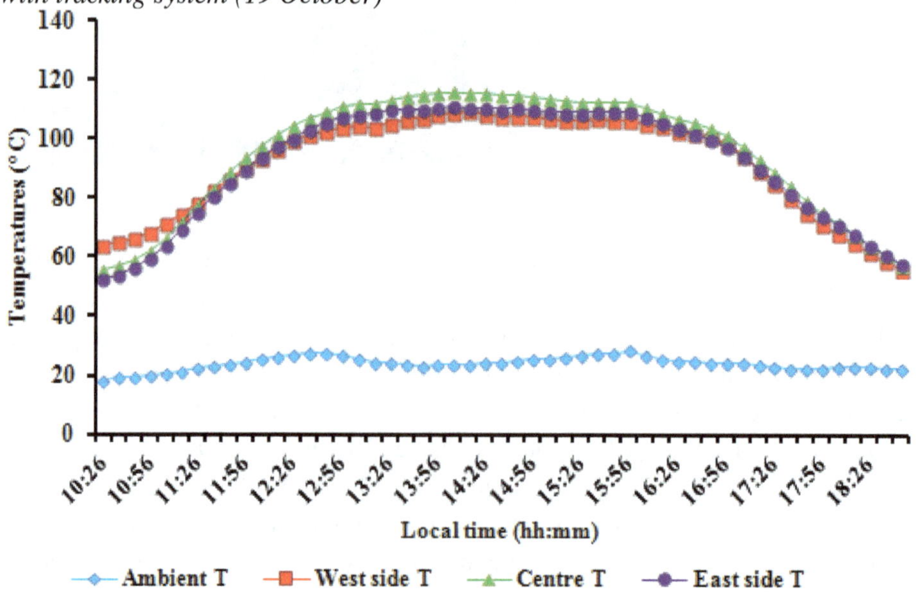

The absorber plate exchanges heat by convection with the internal air of the solar cooker. Therefore, the temperature uniformity of the absorber plate surface resulting from the use of the tracking system can significantly increase the temperature of the internal air and then improve the solar cooker performances.

6.2.5. Temperature Profile for the Stagnation Test With an Internal Reflector

Figure 20 displays characteristic fluctuations in solar irradiation on a horizontal surface, ambient temperature, absorber plate temperature, and inside air temperature of the cooker. These parameters were recorded during a stagnation test under clear sky conditions, in absence of wind.

Figure 20. Diurnal variation of solar irradiation, ambient temperature, measured plate temperature and inside air temperature of the cooker under stagnation test condition with an internal reflector (30 October)

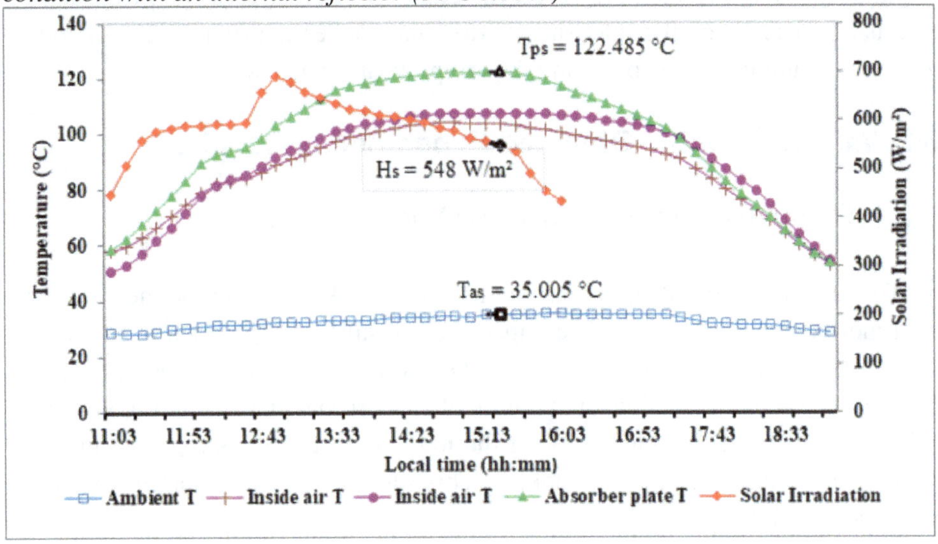

The highest temperature reached by the absorber plate was 122.4 °C, recorded at 15:13 local time (see Figure 20). This occurred under solar irradiation of 557 W/m^2 and an ambient temperature of 34 °C. The first figure of merit F_1 was calculated as 0.157 m^2 K/W using equation (3). Notably, this calculated figure of merit is considerably higher than the one obtained in a test without the use of an internal reflector.

Comparing this test with the one conducted without any reflector (refer to Figure 9), it is observed that the stagnation temperature of the absorber plate is lower in the case with an internal reflector, measuring 119.5 °C compared to 122.4 °C. The summarized values of the stagnation test for both cases are presented in table 4 below.

Table 4. Stagnation test values

	Test without reflector	Test with an internal reflector
T_{ps} (°C)	119.5	122.4
T_{as} (°C)	24.7	34
H_s (W/m²)	760	557
T_{ia_max} (°C)	96	107.5
F_1 (m² K/W)	0.124	0.157

Despite the solar irradiation at the time of stagnation is lower for the test with an internal reflector, the first figure of merit is much higher compared with the test without any reflector. This is probably due to the radiation reflected on the absorber plate from the reflector. However, the addition of other reflectors (internal or external) especially on the side walls of the solar cooker can significantly increase the temperature inside the oven and thus improve its efficiency.

6.3. Validation of the Obtained Experimental Results

6.3.1. Comparison of Results With Other Cookers

To assess the thermal performance of the box cooker with an inclined aperture area more comprehensively, we compared the results of experimental stagnation tests from our cooker with those of other box cooker prototypes worldwide. Table 5 indicates that the stagnation temperature of the absorber plate in our implemented solar cooker (119.5 °C) ranks among the highest temperatures reported in other studies involving different box cookers. This suggests a significant enhancement in the achieved design performance.

Table 5. Comparison of the stagnation temperatures of our solar cooker with other box cookers made around the world

Reference	Date of test	H_s (W/m²)	T_{as} (°C)	T_{ps} (°C)	Time at Stagnation
Sethi et al., 2014	01/2010	980	20	125	13:00
Yettou et al., 2014	05/02/2013	792	12.5	127.6	13:05
Nahar, 1990	-	760	28	132	14:00

continued on following page

Table 5. Continued

Reference	Date of test	H_s (W/m²)	T_{as} (°C)	T_{ps} (°C)	Time at Stagnation
Kumar et al., 2010		858	38	138	13:40
Vaishya et al., 1985	23/09/1982	1020	32	122	12:00
El-Sebaii & Ibrahim, 2005	21/07/2002	-	-	122	-
Harmim et al., 2010	23/07/2008	960	48	140	12:00
Mahavar et al., 2012	17/06/2009	945	35	140	13:00
Misra & Aseri, 2012	01/01/2010	662	39.6	133.9	11:30
Abed, 2011	04/03/2011	742.5	36.5	130	11:55
Farooqui, 2013	13/11/2012	700	25	110	11:55
Harmim et al., 2013	30/12/2012	719	18.2	127.7	12:30
Present study	26/09	760	24.7	119.5	13:30

The first and second figures of merit values are compared with those of other box-type cookers, as presented in Table 6. During testing in the autumn season at the Agadir site, without the incorporation of additional reflectors, our cooker design demonstrates notable values for both F_1 and F_2. This can be attributed to effective thermal insulation and efficient reception of solar rays.

Table 6. Comparison of the values of the first and second figure of merit of our solar cooker with other box cookers made around the world

Reference	F_1 (m² K/W)	F_2	Water amount (kg)
Sethi et al., 2014	0.16	0.54	-
Yettou et al., 2014	0.145	0.391	1.6
El-Sebaii & Ibrahim, 2005	0.15	0.407	4
Mahavar et al., 2012	0.116	0.466	1.2
Misra & Aseri, 2012	0.1424	0.408	1
Abed, 2011	0.11	0.34	-
Farooqui, 2013	0.1258	0.369	1.37
Harmim et al., 2013	0.152	0.47	3.5
Present study	0.124	0.436	3

The adjusted cooking power is additionally compared with values from other solar box cookers in Table 7. Following the European Standard suggested by Funk, our solar cooker is distinguished by effective thermal insulation, indicated by a low slope of the cooking power regression line, and a relatively large receiving surface, reflected in the initial cooking power.

Table 7. Adjusted cooking power of our solar cooker compared with other box cookers made around the world

Reference	Adjusted cooking power	Standardized power (W)
Funk, 2000	$P_{adj} = 125 - 1.58\ \Delta T$	46
Sethi et al., 2014	$P_{adj\_h} = -14.923 + 2.4418\ \Delta T$	103
Sethi et al., 2014	$P_{adj\_30°} = -23.43 + 3.9976\ \Delta T$	175
Yettou et al., 2014	$P_{adj} = 96.706 - 0.9611\ \Delta T$	48
Yettou et al., 2014	$Padj = 110 - 1.6715\ \Delta T$	26
El-Sebaii & Ibrahim, 2005	$P_{adj} = 103.92 - 1.598\ \Delta T$	24
Mahavar et al., 2012	$P_{adj} = 103.5 - 1.474\ \Delta T$	30
Abed, 2011	$P_{adj} = 5.321 - 19.532\ \Delta T$	-
Harmim et al., 2013	$P_{adj} = 136.28 - 1.142\ \Delta T$	78.9
Present study	$P_{adj} = 88.91 - 0.9219\ \Delta T$	41.3
Present study	$P_{adj} = 96.131 - 0.9655\ \Delta T$	47.8
Present study	$P_{adj} = 103.5 - 1.0227\ \Delta T$	52.3

The comparison of the performance of the cooker produced with other cookers performed around the world was also carried out, table 8, in terms of energy and exergy efficiencies following the procedure tested by Ozturk (Ozturk, 2004) and Panwar et al. (Panwar et al., 2010). Our equipment, tested with a load of 3 L of water, is characterized by very acceptable energy efficiency (19.36%) and exergy efficiency (10.12%).

Table 8. Calculated energy and exergy efficiencies parameters of our solar cooker compared with other box cookers made around the world.

	Variables (daily average)				
References	H (W/m²)	E_o (kJ)	Ξ_o (kJ)	ç (%)	ø (%)
Ozturk, 2004	785	21.6	2.50	18.30	2.20
Panwar et al., 2010	-	25.48	1.42	15.45	0.81
Kumar, 2004	816.93	15.42	1.56	26.21	2.85
Yettou et al., 2015	989.40	9.04	0.97	16.23	1.20
Present study	636	31.56	15.83	19.36	10.12

7. CONCLUSION

The purpose of this work was to design, built and test a solar cooker. TRNSYS software was used to model and simulate the operation of the solar cooker through two tools: TRNSYS simulation studio and TRNBuild. The simulation was carried out with different slopes of the aperture area in order to analyze the effect of the inclination angle in the amount of solar radiation received by the solar cooker. It was found that that the optimal angle of inclination of the aperture area is always close to the latitude of the study site. A comparison was performed between the horizontal and inclined solar cookers. We demonstrated that the inclined solar cooker exhibits highest first figure of merit, which indicates good optical efficiency and low heat loss factor. The performances of the designed solar box cooker were evaluated and analyzed according to the International Norms and Standards. The salient features of our study are summarized below:

- The solar cooker is characterized by a good value of the first figure of merit F_1 (0.124 m^2 K / W) due to the good interception of solar rays as higher value of F_1 proves the good optical efficiency of the equipment. The second figure of merit F_2 (0.436) is also high, indicating good heat transfer efficiency from the absorber plate and the air inside to the contents (3 kg of water in our case) of the cooking vessels.
- The standardized cooking power with a temperature difference of 50 °C was 41.3 W and 52.3 W for the tests using single and two cooking vessels respectively.
- The cooker tested with a load of 3 L of water is characterized by very acceptable energy efficiency (19.36%) and exergy efficiency (10.12%). These values are among the best values compared to those published by other authors.
- A tracking system has been developed with the goal of maintaining the solar cooker area oriented towards the sun as it traverses the sky daily. Additionally, an internal reflector has been incorporated to enhance the solar radiation received by the absorber plate. The test results indicate an improvement in the efficiency of the solar cooker. Furthermore, the inclusion of additional reflectors, such as external reflectors, can substantially enhance overall efficiency.

ACKNOWLEDGMENT

This study was supported by The Moroccan Ministry of Higher Education and Research in the framework of (PPR/2015/31) project.

REFERENCES

Abed, F. (2011). Experimental investigation of thermal performance of solar cooker with reflector. *European Journal of Scientific Research*, *56*, 112–123.

Algifri, A. H., & Al-Towaie, H. A. (2001). Efficient orientation impacts of box-type solar cooker on the cooker performance. *Solar Energy*, *70*(2), 165–170. DOI: 10.1016/s0038-092x(00)00136-5

Bejan, A. (1987). Unification of three different theories concerning the ideal conversion of enclosed radiation. *Journal of Solar Energy Engineering*, *109*(1), 46–51. DOI: 10.1115/1.3268177

Coccia, G., Di Nicola, G., Pierantozzi, M., Tomassetti, S., & Aquilanti, A. (2017). Design, manufacturing, and test of a high concentration ratio solar box cooker with multiple reflectors. *Solar Energy*, *155*, 781–792. DOI: 10.1016/j.solener.2017.07.020

Cuce, E., & Cuce, P. M. (2013). A comprehensive review on solar cookers. *Applied Energy*, *102*, 1399–1421. DOI: 10.1016/j.apenergy.2012.09.002

Duffie, J. A., & Beckman, W. A. (2013). *Solar engineering of thermal processes* (4th ed.). John Wiley and Sons.

El-Sebaii, A. A., & Ibrahim, A. (2005). Experimental testing of a box-type solar cooker using the standard procedure of cooking power. *Renewable Energy*, *30*(12), 1861–1871. DOI: 10.1016/j.renene.2005.01.007

Farooqui, S. Z. (2013). A gravity based tracking system for box type solar cookers. *Solar Energy*, *92*, 62–68. DOI: 10.1016/j.solener.2013.02.024

Funk, P. A. (2000). Evaluating the international standard procedure for testing solar cookers and reporting performance. *Solar Energy*, *68*(1), 1–7. DOI: 10.1016/s0038-092x(99)00059-6

Ghosh, S. S., Biswas, P. K., & Neogi, S. (2017). Thermal performance of solar cooker with special cover glass of low-e antimony doped indium oxide (IAO) coating. *Applied Thermal Engineering*, *113*, 103–111. DOI: 10.1016/j.applthermaleng.2016.10.185

Harmim, A., Belhamel, M., Boukar, M., & Amar, M. (2010). Experimental investigation of a box-type solar cooker with a finned absorber plate. *Energy*, *35*(9), 3799–3802. DOI: 10.1016/j.energy.2010.05.032

Harmim, A., Merzouk, M., Boukar, M., & Amar, M. (2013). Design and experimental testing of an innovative building-integrated box type solar cooker. *Solar Energy*, *98*, 422–433. DOI: 10.1016/j.solener.2013.09.019

Hernández-Luna, G., & Huelsz, G. (2008). A solar oven for intertropical zones: Evaluation of the cooking process. *Energy Conversion and Management*, *49*(12), 3622–3626. DOI: 10.1016/j.enconman.2008.07.012

Heywood, H. (1971). Operating experiences with solar water heating. *Journal of the Institute of Heating and Ventilating Engineers*, *39*, 63–69. DOI: 10.4236/oalib.1105445

Jebaraj, S., & Srinivasa, P. R. (2015). High-efficiency solar oven for tropical countries. *Journal of Engineering and Applied Sciences (Asian Research Publishing Network)*, *10*, 10213–10217.

Kotas, T. J. (1985). Basic exergy concepts. In *The Exergy Method of Thermal Plant Analysis* (pp. 29–56). https://doi.org/DOI: 10.1016/b978-0-408-01350-5.50009-x

Krane, R. J. (1987). A second law analysis of the optimum design and operation of thermal energy storage systems. *International Journal of Heat and Mass Transfer*, *30*(1), 43–57. DOI: 10.1016/0017-9310(87)90059-7

KREITH. F., KREIDER, J. F., & GOSWAMI, D. Y. (2000). *Principles of solar engineering*. Taylor & Francis.

Kumar, N., Chavda, T., & Mistry, H. N. (2010). A truncated pyramid non-tracking type multipurpose domestic solar cooker/hot water system. *Applied Energy*, *87*(2), 471–477. DOI: 10.1016/j.apenergy.2009.06.031

Kumar, S. (2004). Thermal performance study of box type solar cooker from heating characteristic curves. *Energy Conversion and Management*, *45*(1), 127–139. DOI: 10.1016/s0196-8904(03)00103-1

Mahavar, S., Sengar, N., Rajawat, P., Verma, M., & Dashora, P. (2012). Design development and performance studies of a novel single family solar cooker. *Renewable Energy*, *47*, 67–76. DOI: 10.1016/j.renene.2012.04.013

Mirdha, U. S., & Dhariwal, S. R. (2008). Design optimization of solar cooker. *Renewable Energy*, *33*(3), 530–544. DOI: 10.1016/j.renene.2007.04.009

Mishra, R. S., & Sabberwal Prakash, S. P. (1984). Evaluation of solar cooker thermal performance using different insulating materials. *International Journal of Energy Research*, *8*(4), 393–396. DOI: 10.1002/er.4440080410

Misra, R., & Aseri, T. K. (2012). Thermal performance enhancement of box-type solar cooker: A new approach. *International Journal of Sustainable Energy*, *31*(2), 107–118. DOI: 10.1080/1478646x.2011.552978

MUJAHID, A. M. (1994). Optimum tilt angle for solar collection systems. *International Journal of Solar Energy*, *14*(4), 191–202. DOI: 10.1080/01425919408909810

Mullick, S. C., Kandpal, T. C., & Saxena, A. K. (1987). Thermal test procedure for box-type solar cookers. *Solar Energy*, *39*(4), 353–360. DOI: 10.1016/s0038-092x(87)80021-x

Nahar, N. M. (1990). Performance and testing of an improved hot box solar cooker. *Energy Conversion and Management*, *30*(1), 9–16. DOI: 10.1016/0196-8904(90)90051-y

Nahar, N. M., Marshall, R. H., & Brinkworth, B. J. (1994). Studies on a hot box solar cooker with transparent insulation materials. *Energy Conversion and Management*, *35*(9), 787–791. DOI: 10.1016/0196-8904(94)90107-4

Nkhonjera, L., Bello-Ochende, T., John, G., & King'ondu, C. K. (2017). A review of thermal energy storage designs, heat storage materials and cooking performance of solar cookers with heat storage. *Renewable & Sustainable Energy Reviews*, *75*, 157–167. DOI: 10.1016/j.rser.2016.10.059

Ozturk, H. H. (2004). Energy and exergy efficiencies of a solar box-cooker. *International Journal of Exergy*, *1*(2), 202. DOI: 10.1504/ijex.2004.005091

Panwar, N. L., Kothari, S., & Kaushik, S. C. (2010). Experimental investigation of energy and exergy efficiency of masonry-type solar cooker for animal feed. *International Journal of Sustainable Energy*, *29*(3), 178–184. DOI: 10.1080/14786461003628679

Petela, R. (2003). Exergy of undiluted thermal radiation. *Solar Energy*, *74*(6), 469–488. DOI: 10.1016/s0038-092x(03)00226-3

Rehfuess, E. (2006). *Fuel for life: Household energy and health*. World Health Organization.

Saxena, A., Varun, Pandey, S. P., & Srivastav, G. (2011). A thermodynamic review on solar box type cookers. *Renewable & Sustainable Energy Reviews*, *15*(6), 3301–3318. DOI: 10.1016/j.rser.2011.04.017

Sethi, V. P., Pal, D. S., & Sumathy, K. (2014). Performance evaluation and solar radiation capture of optimally inclined box type solar cooker with parallelepiped cooking vessel design. *Energy Conversion and Management, 81*, 231–241. DOI: 10.1016/j.enconman.2014.02.041

Soria-Verdugo, A. (2015). Experimental analysis and simulation of the performance of a box-type solar cooker. *Energy for Sustainable Development, 29*, 65–71. DOI: 10.1016/j.esd.2015.09.006

Terres, H., Lizardi, A., Chávez, S., López, R., & Vaca, M. (2017). Exergy applied to the heat conduction analysis in glass covers of a solar cooker box-type with internal and external reflectors. *Journal of Physics: Conference Series, 792*, 012012. DOI: 10.1088/1742-6596/792/1/012012

Terres, H., Lizardi, A., López, R., Vaca, M., & Chávez, S. (2014). Mathematical model to study solar cookers box-type with internal reflectors. *Energy Procedia, 57*, 1583–1592. DOI: 10.1016/j.egypro.2014.10.150

Todd, J. J., & Miller, S. (1999). *Performance testing of cardboard, solar box-cookers*. Graham Thorpe.

Toonen, H. M. (2009). Adapting to an innovation: Solar cooking in the urban households of Ouagadougou (Burkina Faso). *Physics and Chemistry of the Earth Parts A/B/C, 34*(1–2), 65–71. DOI: 10.1016/j.pce.2008.03.006

TRNSYS. (1979). *A transient simulation program*. Solar Energy Laboratory, University of Wisconsin.

Vaishya, J. S., Tripathi, T. C., Singh, D., Bhawalkar, R. H., & Hegde, M. S. (1985). A hot box solar cooker: Performance analysis and testing. *Energy Conversion and Management, 25*(3), 373–379. DOI: 10.1016/0196-8904(85)90057-3

Yettou, F., Azoui, B., Ali, M., Panwar, N. L., Gama, A., & Toufik, A. (2014). Performance parameters determination by experimental testing of box solar cooker with south tilted collecting surface.

Yettou, F., Azoui, B., Malek, A., Panwar, N. L., & Gama, A. (2015). Energetic and exergetic evaluation of solar box cooker in Algerian climatic conditions. *International Journal of Exergy, 16*(3), 337. DOI: 10.1504/ijex.2015.068230

Yettou, F., Azoui, B., Malek, A., Panwar, N. L., Gama, A., Arrif, T., & Merarda, H. (2023). Comparative assessment of two different designs of box solar cookers under Algerian Sahara conditions. *Journal of Renewable Energies, 18*(2). Advance online publication. DOI: 10.54966/jreen.v18i2.500

APPENDIX

Nomenclature

Symbol Description Unit

A_p Solar box cooker aperture area m^2
C Specific heat capacity J/kg. K
E_i Energy input kJ
E_o Energy output kJ
F_1 First figure of merit m^2.K/W
F_2 Second figure of merit [-]
G_{av} Average solar irradiance W/m^2
G_{ref} Reference illumination level equal to 700 W/m^2 W/m^2
H_s Solar radiation on a horizontal surface W/m^2
$\overline{H}_s$ Average horizontal solar radiation over Δt W/m^2
M Mass kg
P Cooking power Watt
Ps Standard cooking power Watt
T_0 Reference Temperature K
T_a Ambient Temperature K
$\overline{T}_a$ Average ambient temperature over Δt °C
T_{as} Ambient Stagnation Temperature °C
T_{boil} Boiling Temperature °C
T_{boil} Boiling time min
T_d Temperature difference °C
T_{ps} Absorber plate Stagnation Temperature °C
T_s Surface Temperature of the Sun K
T_w Water Temperature K
ΔT Temperature difference °C
Δt Time interval min

Greek Symbols

η Energy efficiency %
Ξ_i Exergy input kJ
Ξ_o Exergy output kJ
ϵ Thermal exergy kJ
ϵ_w Thermal exergy content of water kJ

ψ Exergy efficiency %

Chapter 3

Innovation and Digitization of Solar Cookers Towards Sustainable Energy Management

Mohammed Hmich
https://orcid.org/0009-0007-2880-5766
Faculty of Science, Mohammed First University, Oujda, Morocco

Bilal Zoukarh
Faculty of Science, Mohammed First University, Oujda, Morocco

Sara Chadli
Faculty of Science, Mohammed First University, Oujda, Morocco

Rachid Malek
https://orcid.org/0000-0002-8304-3254
ENSA, Mohammed First University, Oujda, Morocco

Olivier Deblecker
https://orcid.org/0000-0003-2156-9450
Faculty of Engineering, University of Mons, Belgium

Khalil Kassmi
https://orcid.org/0000-0002-4808-873X
Faculty of Science, Mohammed First University, Oujda, Morocco

Najib Bachiri
Association Humain and Environnement of Berkane, Morocco

ABSTRACT

This chapter presents the digitization and supervision of a photovoltaic solar cooker using a Raspberry Pi Pico W microcontroller. The system collects meteorological and electrical parameters (voltage, current, power, efficiency, temperatures) in

DOI: 10.4018/979-8-3373-2535-4.ch003

real time and automatically regulates the cooker's operation. Data, transmitted via the MQTT protocol, are instantly displayed on a Tkinter interface and archived in MySQL databases to ensure full traceability. The results show that the developed architecture enables: Real-time reception and display of data, Timestamped storage for individualized monitoring, Deferred visualization through an interactive web application based on Dash. This centralized and scalable solution provides a reliable tool for supervision, performance analysis, and adaptation of solar cookers to various contexts.

This chapter presents the digitization and supervision of a photovoltaic solar cooker using a Raspberry Pi Pico W microcontroller. The system collects meteorological and electrical parameters (voltage, current, power, efficiency, temperatures) in real time and automatically regulates the cooker's operation. Data, transmitted via the MQTT protocol, are instantly displayed on a Tkinter interface and archived in MySQL databases to ensure full traceability. The results show that the developed architecture enables: Real-time reception and display of data, Timestamped storage for individualized monitoring, Deferred visualization through an interactive web application based on Dash. This centralized and scalable solution provides a reliable tool for supervision, performance analysis, and adaptation of solar cookers to various contexts.

I INTRODUCTION

The energy transition still faces many shortcomings, particularly in key sectors such as cooking. Indeed, many communities continue to rely on fossil fuels (gas, oil, etc.), while others use cooking methods that lead to uncontrolled forest degradation (Imani & Moore-Delate, n.d.; Marín et al., 2021; Yolcan, 2023). Against this backdrop, our project aims to introduce a sustainable and environmentally friendly solution, harnessing photovoltaic solar energy and battery storage. This approach presents itself as a reliable alternative, particularly suited to tropical countries, which benefit from high solar intensity (Mawire et al., 2024; Simon Prabu et al., 2023; Aquilanti et al., 2023; Lecuona-Neumann et al., 2024). To validate the efficiency of our cooker, laboratory tests are essential. However, to ensure full validation and daily performance monitoring, it is crucial to install these cookers in households. These installations will enable the technology to be tested in real-life conditions, allowing performance to be monitored remotely, regardless of geographical location. This field deployment is also a key step towards reducing the ecological impact of using wood as an energy source for cooking. The Raspberry Pi Pico W plays a central role in the operation of our cooker, acting as the device's brain. It manages all the tasks involved in controlling the system, such as choosing the cooking mode, calculating electrical and thermal parameters, and displaying data on the LCD screen. In addition, we need to add an essential functionality: sending data via a communication protocol.

The MQTT (Message Queuing Telemetry Transport) protocol was chosen for this purpose. Very popular in the field of the Internet of Things (IoT), MQTT is a lightweight messaging protocol designed to facilitate communication between devices connected over low-bandwidth or unreliable networks (Rodríguez-Muñoz et al., 2025; Zeghida et al., 2024; Lakshminarayana & Santhi Thilagam, 2025; Das & Jain, 2025; Zeghida et al., 2025). This solution is ideal for our project, given that we will be installing the cookers in a village where network quality could be variable. MQTT also enables us to manage several cookers, each with its own "topic" (string used to identify messages and organize communication). This feature will enable us to monitor and control each cooker independently, while guaranteeing reliable and efficient data exchange. The data collected is sent to the management center in our laboratory, where it is stored in an organized database. An application has already been set up to display the various cookers and their associated electrical and thermal data. This application enables real-time monitoring, providing an overview of the status of each cooker and facilitating management of the information needed to optimize performance and continuously improve the technology.

In the framework of the international cooperation projects WBI 3.3 and LEAP-RE 'SoCoNexGen', we propose the digitalization of an innovative solar cooker powered by photovoltaic solar energy, with the objective of enhancing its reliability, enabling real-time monitoring (both locally and remotely), and collecting instantaneous usage data from household deployment. This data-driven approach will support continuous optimization and cost reduction, facilitating broader dissemination.

This chapter describes the various methods and strategies we have put in place to test how remote data transfer works. This includes the practical steps and techniques we used to check whether the cooker can correctly and reliably send the electrical and thermal data collected to our management center located in the laboratory, using the MQTT communication protocol. We will also detail the tests carried out to ensure that the communication system operates effectively, even in environments with limited network connectivity, as may be the case in rural areas.

II SOLAR PHOTOVOLTAIC COOKER WITH BATTERY STORAGE

II.1 Cooker Overview

As illustrated in the block diagram in Figure 1, prototype no. 4 of our photovoltaic solar cooker — with a total power output of 1.2 kW — is based on a dual power supply architecture, offering flexible and autonomous operation under varying solar conditions. Each power source is managed by an optimized DC/DC boost converter, and the system is composed of the following main functional blocks:

- The first heating element is powered directly by two 600 W photovoltaic panels via a dedicated DC/DC boost converter. This converter is designed to maximize the energy drawn from the solar panels and operates in two distinct modes:

  1. Automatic mode, where the duty cycle is dynamically regulated using the Perturb and Observe (P&O) algorithm to track the maximum power point;
  2. Manual mode, where the user can adjust the duty cycle manually based on specific requirements.

- The second heating element is supplied by a 48 V battery pack, composed of four batteries connected in series, through a second DC/DC boost converter. This converter is optimized to minimize energy losses and functions exclusively in manual mode. The battery pack is independently recharged by another set of two 600 W photovoltaic panels, ensuring continuous operation during low irradiance periods or at night.
- The entire power flow and operational logic are coordinated by a central control box, which plays a critical role in regulating energy distribution. It ensures smooth switching between power sources (solar panels and batteries), manages the boost converters, and maintains overall system stability and safety.

At the heart of the control system is a Raspberry Pi Pico W microcontroller, which supports the digitalization of the cooker. Its main functions include real-time monitoring of electrical and thermal parameters, automatic or manual control of the heating resistors, data display on an integrated LCD screen, and local and remote supervision using the MQTT communication protocol developed as part of this work.

This digital layer enables continuous performance tracking both on-site and remotely through the transmission of data to a centralized monitoring platform. This functionality is essential for evaluating real-world usage, optimizing the system, and supporting broader deployment strategies.

Figure 1. Functional architecture of the proposed solar cooking system

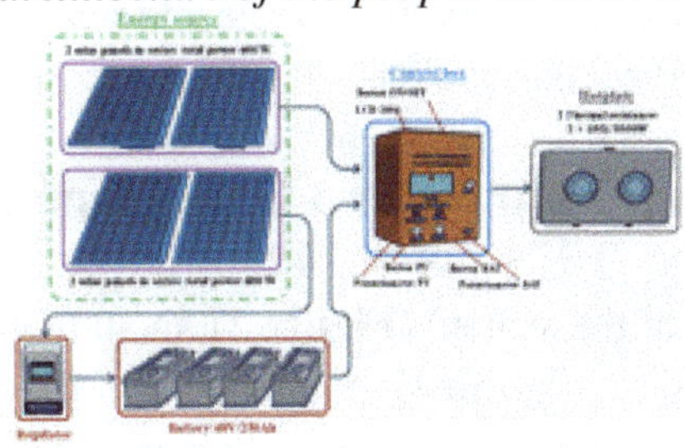

II.2 Digital Control and Languages Used

The solar cooker is controlled by a two-level digital architecture: the on-board microcontroller (Raspberry Pi Pico W) and the LETSER laboratory supervision center. Each level uses a programming language adapted to its constraints and objectives:

- **Raspberry Pi Pico W level:** The Raspberry Pi Pico W acts as the core of the embedded system and is programmed in MicroPython a lightweight version of Python designed specifically for microcontrollers. This choice offers simplicity, hardware compatibility, and low resource consumption. MicroPython efficiently handles sensor readings (voltage, current, temperature), controls relays and heating elements, and manages data transmission via Wi-Fi or temporary access points. Its integrated libraries simplify communication with peripherals using protocols such as I2C, SPI, and UART, enabling rapid integration of devices like LCD displays, digital sensors, and diagnostic interfaces. This makes MicroPython an ideal environment for the on-board development of our solar cooker.
- **Supervision center (LETSER laboratory) level:** Data transmitted through the MQTT protocol is received by a computer running a Python program. Python's readability, modularity, and rich ecosystem make it well suited for this role. The program's responsibilities include receiving MQTT messages, processing data in real time through filtering, conversion, and validation, and dynamically visualizing the information using powerful libraries such as Plotly and Dash. Simultaneously, the data is stored in a structured way within a relational database.

- **Database management:** We chose MySQL due to its robustness and broad compatibility with analysis tools. Using the MySQL Workbench interface, we designed relational schemas and managed tables corresponding to families, weather stations, and cookers, enabling flexible querying of the data. This centralized system provides sustainable and scalable management of the data collected from the solar cookers.

II.3 Overall Operating Diagram and Data Transmission

The overall operation of the system (**Figure 2**), including initial configuration, network connection and data transmission, is shown in the figures below. These illustrate both the local operation of the solar cooker and its integration into a large-scale cloud architecture. When first commissioned, or if no Wi-Fi configuration is available, the Raspberry Pi Pico W automatically activates a temporary Wi-Fi access point. The user can then connect to it via a smartphone or computer, access a local web interface and enter the Wi-Fi network parameters. These settings are then saved in a JSON file.

At each subsequent start-up, if this configuration file is present and valid, the Pico W automatically connects to the Wi-Fi network without user intervention. Once the connection is established, it publishes the collected measurements (voltage, current, temperature) in real time via the MQTT protocol to a remote broker (HiveMQ). This architecture enables autonomous, reliable and secure operation, while offering a simple, intuitive user interface suitable for both local and remote use.

Figure 2. Wi-Fi communication diagram between the smart solar cooker and the supervision server via MQTT

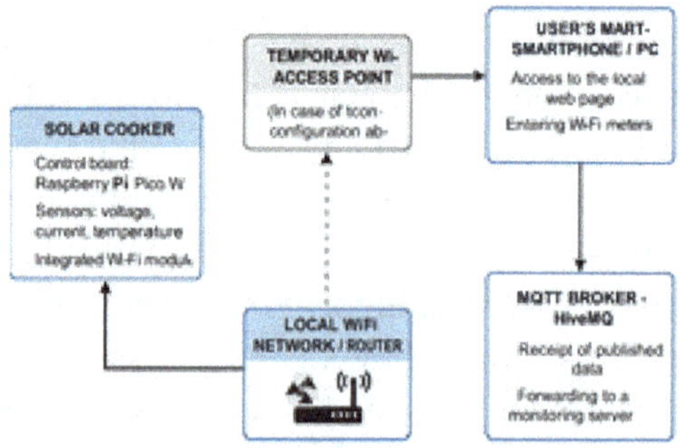

On a larger scale, several photovoltaic cookers installed in families (Family 1 to Family *n*) are connected to the Internet and transmit their data via MQTT to the

HiveMQ broker in the cloud (**Figure 3**). The LETSER laboratory supervision center receives this data in real time, then stores it in a MySQL database.

Two separate databases are maintained:

- One for the cookers, with one table per family,
- The other for the weather station, integrated into the same supervision system.

The data is then displayed in real time on a local screen, and also made accessible via a Dash application hosted on a server. This interface enables several authorized users to consult the performance of the cookers via their web browser, from the server's IP address.

This modular system enables extensive, scalable and interactive supervision, with precise traceability of the energy and thermal performance of each installation.

Figure 3. Communication diagram between domestic solar cookers and the supervision center

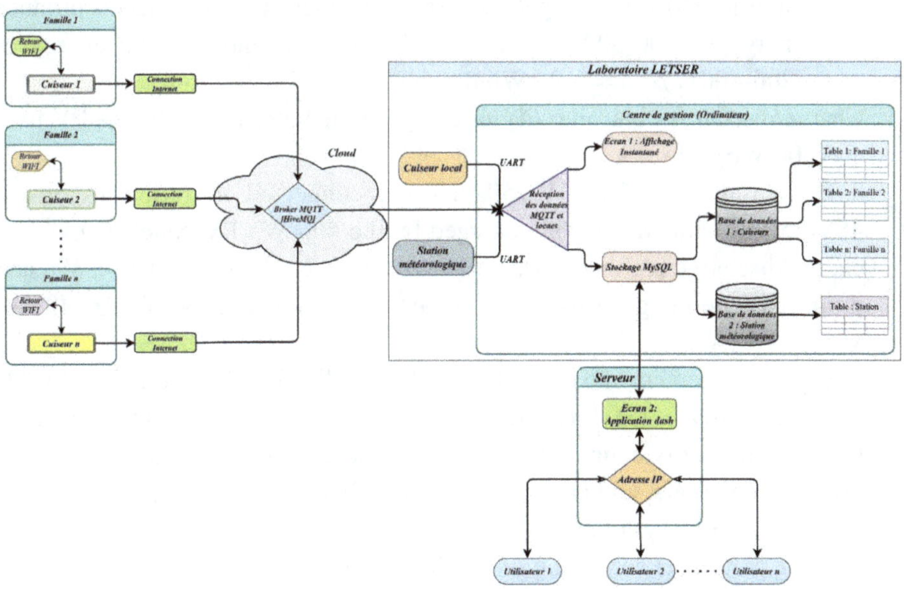

III EXPERIMENTAL VALIDATION OF A SOLAR COOKER DATA COLLECTION AND TRANSMISSION SYSTEM

III.1 Connecting and Starting Up the Solar Cooker

Figure 4 shows the equipment package for cooker 3 (Prototype P4) set up in the LETER laboratory at the University of Oujda (Morocco). The proposed cooker is designed to produce a total energy output of 2-4.5 kWh/d. The cooker's equipment and the various items of equipment required for its testing are:

- Photovoltaic (PV) panels with a total power of 1200 Wp:
  - ✓ Two PV panels **(1)** in series with a total output of 600 Wp, connected directly to the control box **(10)**. They produce energy of 500-600Wp) x7h=3.5-4.2 kWh/day, and cooking energy of (200-450W)x5h = 1-2.5 kWh/d.
  - ✓ Two PV panels **(2)** in series, rated at 600 Wp, connected to the solar batteries **(5)** via a charge/discharge controller **(4)**. These panels produce energy of 500-600Wpx7h=3.5-4.2 kWh/day, and cooking energy of (200-400W)x5h = 1-2 kWh/d.
- Charge/discharge controller **(4)** for charging the batteries **(5)** by the PV panels (600 Wp).
- 4 Solar Batteries **(5)**, 12V/250Ah each, connected in series to form a 48V/250Ah battery. They are charged by the 600 Wp PV panels **(2)** via the charge/discharge regulator **(4)**. They can store a total energy of 12 kWh, and produce, depending on user requirements, cooking energy of (200-400W) x5h=1-1.2 kWh/d.
- Energy management and control box **(10)** supplied by **600 Wp** PV panels **(1)** and solar batteries **(5)**. It regulates the heating of the hot plate **(11)**, powered by the two energy sources (PV panels **(1)** and solar batteries **(5)**), via two DC/DC converters, and an electronic board that controls the operation of the complete cooker system.
- Heating plate **(11)** consisting of two heating resistors (left and right), with a power rating of 2 kW and a temperature of 1000°C.
- Meteorological station **(8)** consisting of a Pyranometer **(3)** and thermal sensor to measure ambient temperature illuminance intensity.
- A computer and two screens **(9)** to program the control board (Raspberry Pi microcontroller), control, regulate and monitor the operation of the cooker, both locally and remotely via the Internet **(7)**.

Figure 4. Broaching of the cooker 4 produced in this LAEP-RE 'SoCoNextGen' project.

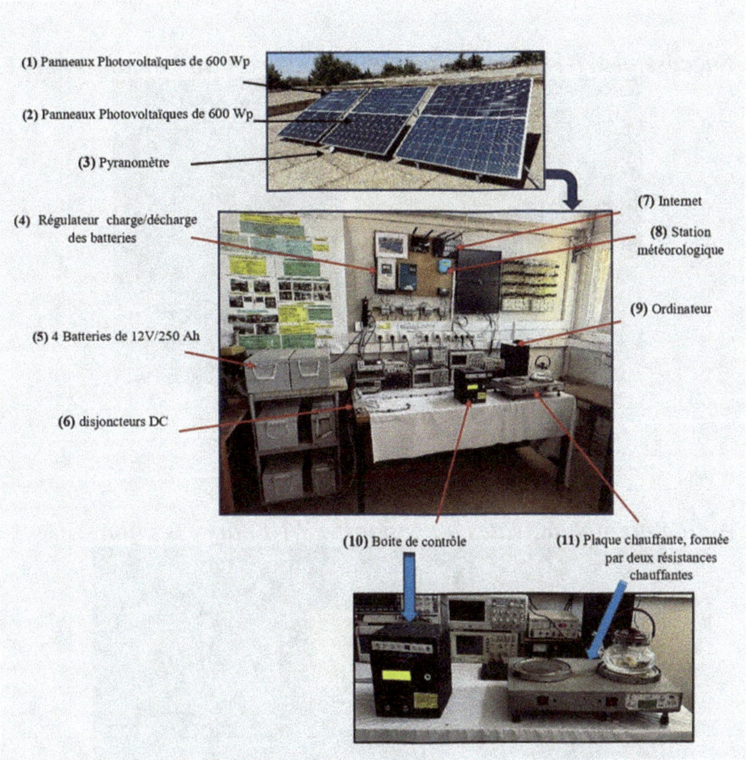

III.2 Internet Connection via Wi-Fi

The **Raspberry Pi Pico W** built into our cooker is equipped with a Wi-Fi module for connection to a local network (home or mobile Wi-Fi). Once connected, the microcontroller can transmit and receive data in real time.

The Wi-Fi connection is configured automatically at start-up via a configuration file stored in the microcontroller's internal memory. This file contains the Wi-Fi network identifiers (**Figure 5**).

If no network configuration is detected at start-up, the cooker automatically activates a temporary Wi-Fi access point ("Access Point" mode) (**Figure 6**). This local network allows the user to connect directly to the cooker via a smartphone, tablet or computer. An embedded web interface then appears, offering a simple form for entering Wi-Fi network parameters (SSID and password) **(Figure 7)**. Once the information has been entered, the device reboots and automatically attempts to connect to the configured network.

Figure 5. Successful Wi-Fi connection between cooker and local router

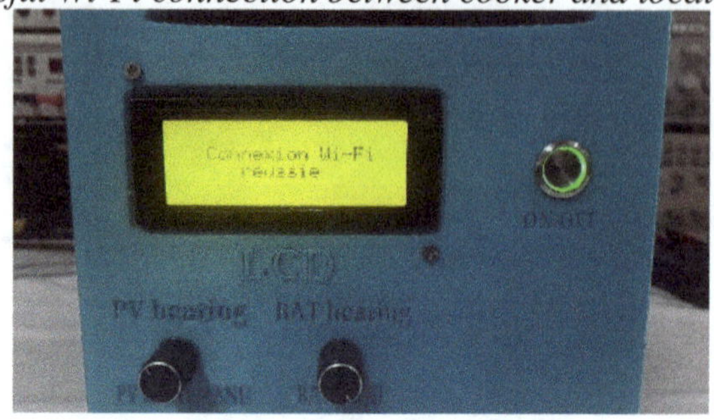

Figure 6. Automatically activates a temporary Wi-Fi access point

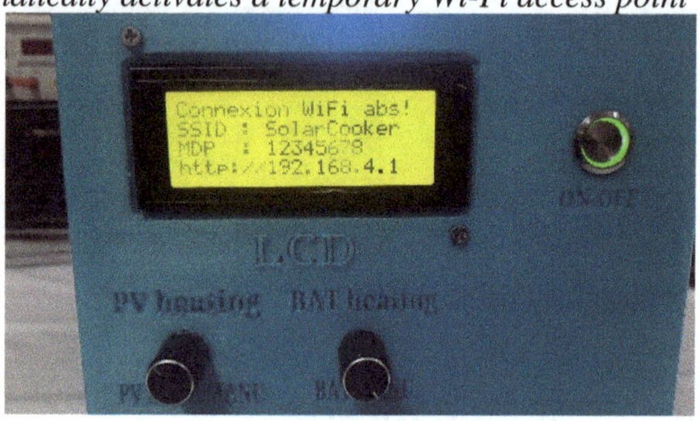

Figure 7. Screenshot of Wi-Fi configuration web page displayed on smartphone

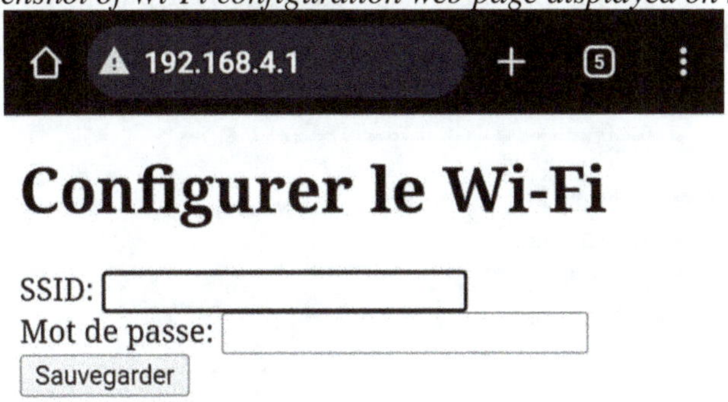

This solution makes the system fully autonomous and easy to deploy, even in environments without technical support, while avoiding the need for manual code modification or physical access to the microcontroller board. What's more, it enhances the device's flexibility by enabling rapid, secure remote configuration, thus reducing the risk of human error and the costs associated with on-site intervention. This approach also enhances system scalability, facilitating the integration of future upgrades without service interruption.

III.3 Data Transmission via MQTT Protocol

Once connected to the Internet, the cooker uses the MQTT (Message Queuing Telemetry Transport) protocol to transmit data to a remote server called a broker (**Figure 8**). For our system, we chose **HiveMQ**, a free, public MQTT broker that enables fast, lightweight and reliable message exchange (**Figure 9**).

Figure 8. MQTT communication between the cooker and the HiveMO broker

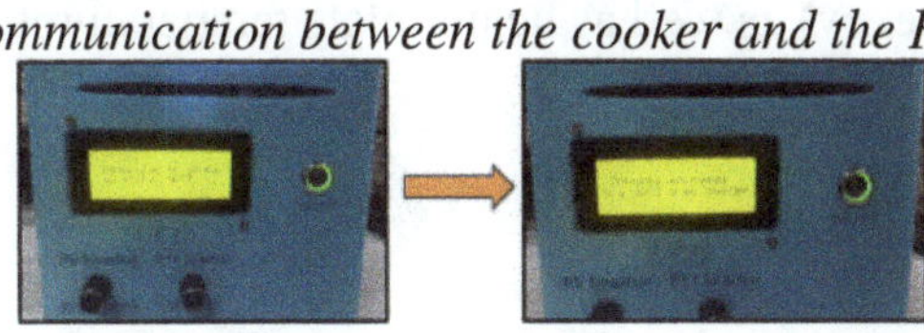

Figure 9. Screenshot of the HiveMO dashboard showing published messages

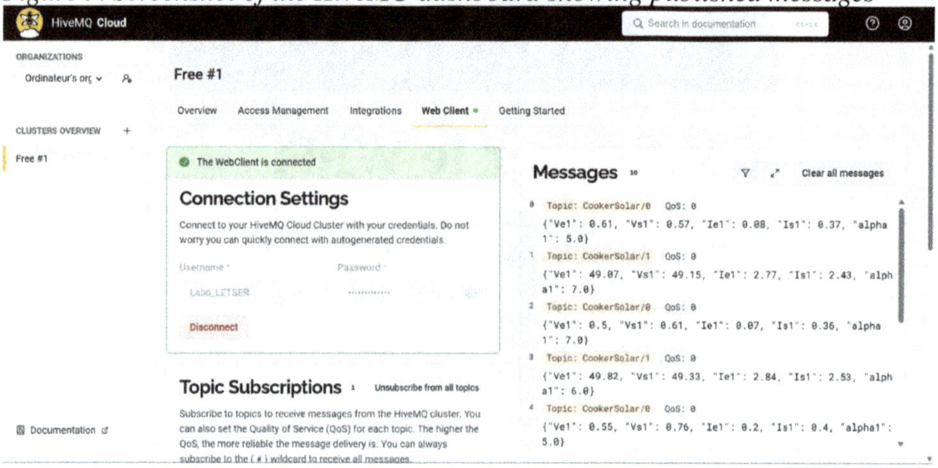

Each cooker publishes its data (power, voltage, temperature, etc.) on a specific, uniquely identified **MQTT topic**. This makes it possible to individually track several cookers in different households.

IV SOLAR COOKER SUPERVISION ARCHITECTURE: RECEPTION, STORAGE AND VISUALIZATION

IV.1 Real-Time Reception and Local Display via Tkinter

Data from the solar cookers, transmitted via the MQTT protocol, are received in real time by our main program. These are immediately displayed on a local graphical interface developed with Tkinter, installed on the laboratory's supervision workstation. This interface enables live monitoring of electrical quantities (voltage, current, power, efficiency) as well as heater plate temperatures (see **Figure 10**).

Figure 10. Photo of the supervision station installed in the laboratory

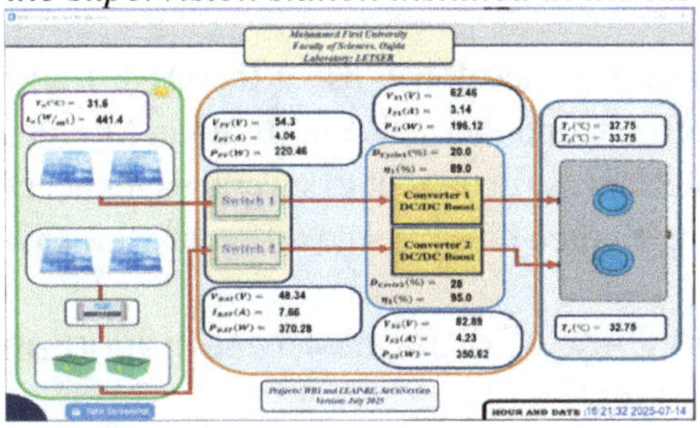

IV.2 Centralized Storage in MySQL Workbench

In parallel with the real-time display, all data received is automatically stored in several MySQL Workbench databases (**Figure 11**), each dedicated to a solar cooker and hosted in our laboratory. Each measurement is recorded with its timestamp, ensuring complete, individualized traceability of each system's operation. This distributed architecture enables fine-tuned, targeted use of the data, whether for analysis, specific reports or maintenance actions.

Figure 11. Screenshot of MySQL databases containing thermal and electrical measurements

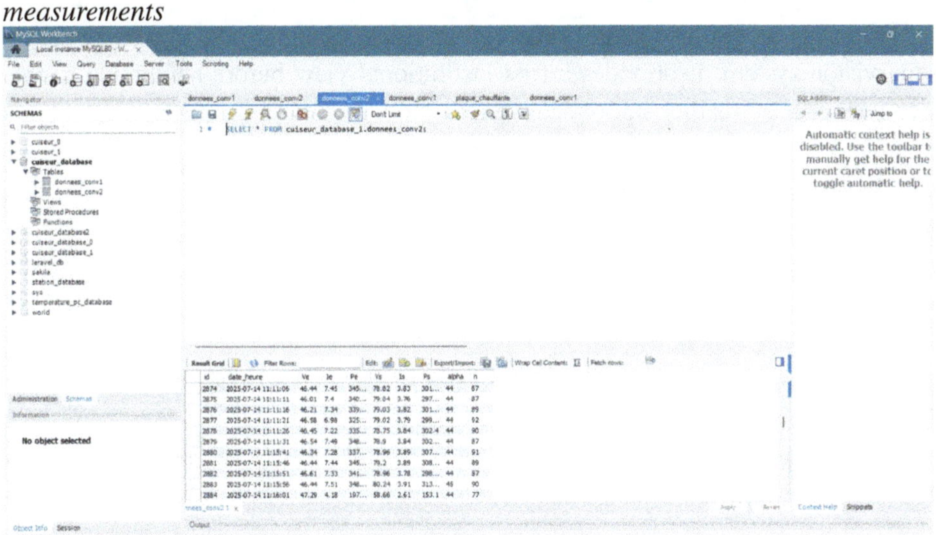

IV.3 Deferred Visualization via Dash Web Application

The web application was developed using the Dash framework to enable deferred viewing of electrical and thermal data stored in MySQL databases. It offers an interactive user interface, structured into several pages accessible via navigation buttons. The following figures illustrate the different views available.

IV.3.1 Dash Home Page: Corporate Presentation and Partnerships

The home page of the Dash web application (**Figure 12**) has been designed to present institutional information related to the project, the scientific context, and the national and international partners involved. It does not yet contain technical graphics, but serves as an explanatory gateway for the user.

Elements displayed include

- Project name, laboratory (LESPRE) and host university (UMP - Oujda);
- Team composition: PhD students, trainees, supervisors;
- Academic and technical partnerships, divided into two categories:
    - *International partners*: European and North African universities and research centers;
    - *National partners*: local institutions, local authorities and associations.
- An image carousel highlights project logos or photos.

This page, accessible via the home-button, serves as an introduction to the entire supervision system. It provides a first institutional view before navigating to the technical pages (weather station, converters, etc.).

Figure 12. Dash home page interface showing project teams and partners.

IV.3.2 Weather Station Page: Temperature and Irradiance

The weather station page (**Figure 13**) displays interactive graphs showing the evolution of ambient temperature and solar irradiance over time. These data come from the UART module connected to the environmental sensors, and are retrieved from the MySQL database dedicated to the station.

Figure 13. Display of weather data via Dash application (temperature and irradiance)

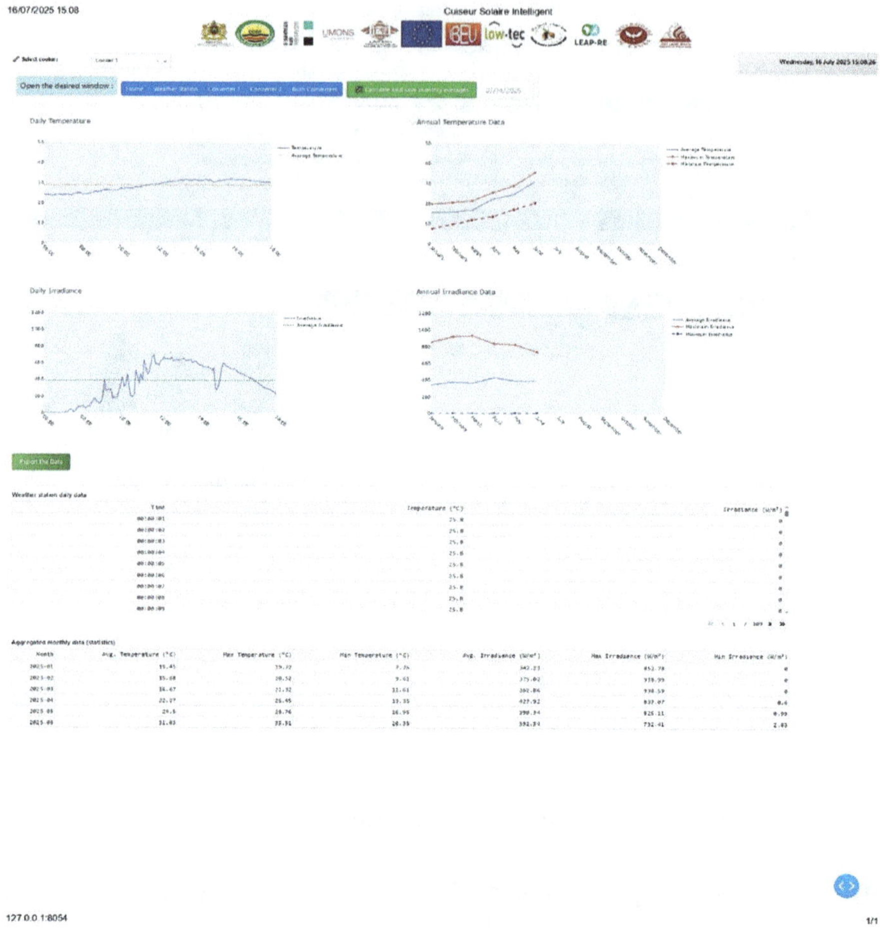

IV.3.3 Converter Page: Delayed Performance Monitoring

This section of the Dash application allows you to view the temporal evolution of electrical and thermal data recorded for a specific converter, based on data stored in MySQL databases.

Interactive Features

The page dedicated to each converter (**Figure 14**) is accessed via buttons (converter1, converter2, two converters) in the navigation menu. It offers the following functions:

- Delayed display of measured quantities: input/output voltage (Ve, Vs), input/output current (Ie, Is), power (Pe, Ps), efficiency η, and duty cycle α ;
- Simultaneous display of associated hotplate temperatures (Tr, Tt, Tc) to analyze the thermal-electrical relationship;
- Interactive graphs grouped by data type (voltage, current, power, etc.) with quick anchor links;
- Dynamic tables for viewing measurements in tabular form;
- Data export in Excel format (Export the Data) for external analysis or archiving.

Figure 14. Delayed monitoring interface for Converter 1: graphical display and measurement table.

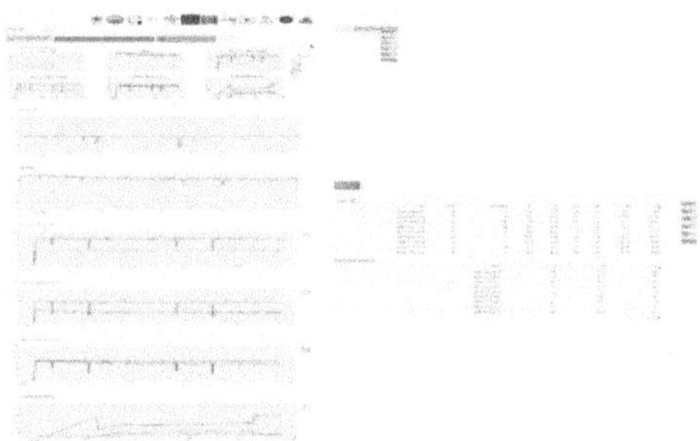

V CONCLUSION

The work presented in this chapter illustrates the relevance of an innovative energy solution, combining photovoltaic solar energy, battery storage and remote digital supervision. By combining efficient hardware technologies (converters, sensors, Raspberry Pi Pico W microcontroller) with lightweight but robust software solutions (MicroPython, MQTT, MySQL, Dash), the solar cooking system developed meets the real needs of rural households, while offering a scalable and adaptable infrastructure. Laboratory tests validated the cooker's thermal performance, while field experiments demonstrated the system's ability to operate reliably in constrained environments, where connectivity and access to electricity are sometimes limited. The choice of the MQTT protocol, combined with an intuitive visualization interface, made it possible to set up real-time data tracking, guaranteeing complete traceability and optimized management of the energy produced and consumed.

This holistic approach shows that it is now possible to design energy systems that are autonomous, intelligent and environmentally friendly. The proposed solar cooker is thus a concrete solution to the challenges of energy transition in rural areas, providing comfort, safety and reduced ecological impact.

ACKNOWLEDGEMENTS

This research is supported within the framework of the projects:

Morocco-Wallonie Brussels Cooperation Program (2023-2027), Wallonie-Bruxelles-International (Belgium), project 3.3

Europe-Africa Research and Innovation call on Renewable Energy LEAP-RE, Project No 963530: SoCoNexGen, Solar Indoor Cooking Systems of the Next Generation, 2022-2025

REFERENCES

Aquilanti, A., Tomassetti, S., Coccia, G., Muccioli, M., & Di Nicola, G. (2023, March). Experimental characterization and performance comparison of four prototypes of panel solar cooker for low to high sun elevations. *Journal of Cleaner Production*, *390*, 136158. DOI: 10.1016/j.jclepro.2023.136158

Das, A., & Jain, R. K. (2025). Performance evaluation of mobile robotic system for monitoring and inspection using MQTT protocol and IoT technique. *International Journal of Intelligent Robotics and Applications*, (June). Advance online publication. DOI: 10.1007/s41315-025-00465-z

Imani, G., & Moore-Delate, E. (n.d.)."Rapport d'étude de la consommation de bois-énergie et des équipements de cuisson de la ville de Kisangani", [Online]. Available at: https://www.cifor-icraf.org/publications/pdf_files/Reports/Imani-GML-Report.pdf

Lakshminarayana, S., & Santhi Thilagam, P. (2025, June). Next-Generation DDoS Attacks on IoT Deployments: Targeting the Advanced Features of MQTT v5.0 Protocol. *IEEE Internet of Things Journal*, *12*(12), 22090–22109. DOI: 10.1109/JIOT.2025.3549784

A. Lecuona-Neumann, J. I. Nogueira-Goriba, A. Famiglietti, M. del C. Rodríguez-Hidalgo, and J. Boubour, "Solar Photovoltaic Cooker with No Electronics or Battery", *Energies*, vol. 17, n(o) 5, Art. n(o) 5, Jan. 2024, doi: .DOI: 10.3390/en17051192

Marín, A. I., Abdul Malak, D., Bastrup-Birk, A., Chirici, G., Barbati, A., & Kleeschulte, S. (2021, September). Mapping forest condition in Europe: Methodological developments in support to forest biodiversity assessments. *Ecological Indicators*, *128*, 107839. DOI: 10.1016/j.ecolind.2021.107839

Mawire, A., Abedigamba, O. P., & Worall, M. (2024, February). Experimental comparison of a DC PV cooker and a parabolic dish solar cooker under variable solar radiation conditions. *Case Studies in Thermal Engineering*, *54*, 103976. DOI: 10.1016/j.csite.2024.103976

Rodríguez-Muñoz, J. D., Tlelo-Cuautle, E., & de la Fraga, L. G. (2025, May). Chaos-based authentication of encrypted images under MQTT for IoT protocol. *Integration (Amsterdam)*, *102*, 102378. DOI: 10.1016/j.vlsi.2025.102378

A. Simon Prabu, V. Chithambaram, M. Anto Bennet, S. Shanmugan, and B. Janarthanan, "Performance of solar cooker with evacuated tubes and photovoltaic panels with phase change materials", *Energy Efficiency*, vol. 16, n(o) 7, p. 72, August 2023, doi: .DOI: 10.1007/s12053-023-10151-3

Yolcan, O.O. (2023). "World energy outlook and state of renewable energy: 10-Year evaluation", *Innovation and Green Development*, vol. 2, n(o) 4, p. 100070, Dec. 2023, doi: .DOI: 10.1016/j.igd.2023.100070

H. Zeghida *et al*, "Enhancing IoT cyber attacks intrusion detection through GAN-based data augmentation and hybrid deep learning models for MQTT network protocol cyber attacks", *Cluster Comput*, vol. 28, n(o) 1, p. 58, Nov. 2024, doi: .DOI: 10.1007/s10586-024-04752-5

H. Zeghida, M. Boulaiche, R. Chikh, A. Patel, A. L. B. Barros, and A. M. Bamhdi, "XMID-MQTT: explaining machine learning-based intrusion detection system for MQTT protocol in IoT environment," *Int. J. Inf. Secur.* vol. 24, n(o) 3, p. 128, May 2025, doi: .DOI: 10.1007/s10207-025-01036-w

Chapter 4
Integration of Thermal Energy Storage Materials for Off-Sunshine Cooking

Navendu Misra
Rajiv Gandhi Institute of Petroleum Technology, India

Atul Sharma
https://orcid.org/0000-0002-3754-8550
Rajiv Gandhi Institute of Petroleum Technology, India

Himanshu Mishra
https://orcid.org/0009-0009-1461-6635
Rajiv Gandhi Institute of Petroleum Technology, India

Saurabh Pandey
https://orcid.org/0000-0001-6946-3642
Rajiv Gandhi Institute of Petroleum Technology, India

ABSTRACT

Thermal Energy Storage (TES) has emerged as a crucial innovation in solar cooking, addressing the major limitation of traditional solar cookers—their dependence on sunlight. This chapter explores the principles, technologies, and applications of TES in solar cookers, emphasizing its transformative role in enabling cooking during off-sunshine hours, such as evenings, nights, and cloudy days. By storing excess thermal energy collected during peak sunlight hours, TES systems extend the operational period of solar cookers, making them more practical, efficient, and reliable. While solar cookers offer an eco-friendly and cost-effective alternative to conventional cooking methods, their reliance on direct sunlight restricts their usability. TES bridges this gap by ensuring uninterrupted cooking and enhancing efficiency. The chapter

DOI: 10.4018/979-8-3373-2535-4.ch004

categorizes TES systems into three main types: sensible heat storage, which utilizes materials like water, oil, or rocks that retain and release heat through temperature changes; latent heat storage, which leverages phase change materials (PCMs) such as paraffin wax or salt hydrates that store and release energy during phase transitions; and thermochemical storage, which employs reversible chemical reactions for long-term heat retention with minimal losses. A comparative analysis highlights their thermal properties, cost effective, and practicality for solar cooking. Effective TES integration requires high energy density, thermal stability, durability, and ease of use, and this chapter discusses how TES systems can be tailored to various solar cooker designs, from small household units to large-scale community systems. The discussion also covers key aspects such as thermal insulation, heat retention, and heat transfer mechanisms to maximize efficiency and minimize heat loss, while safety and user-friendliness are emphasized for adoption in resource-limited settings. TES significantly enhances solar cooking by extending cooking hours, reducing reliance on conventional fuels, and improving user convenience. However, challenges such as heat loss, material degradation, and inefficiencies, along with economic and accessibility barriers, must be addressed. The chapter underscores the need for continued research and innovation to improve TES-integrated solar cookers, making them more affordable, efficient, and accessible for widespread adoption.

1. INTRODUCTION

As the world faces a shortage of conventional energy sources and an alarming rise in greenhouse gas emissions, the need for sustainable energy solutions has become more critical than ever. The primary contributors to this issue are industrialization and rapid population growth, which have significantly increased energy demand. Cooking, a necessity, is predominantly reliant on fossil fuels and electricity, further exacerbating environmental concerns (Zhou et al., 2023).

Renewable energy sources offer a viable solution to both energy shortages and carbon emissions. Among them, solar cooking stands out as an environmentally friendly and sustainable alternative, providing a carbon-free method of food preparation. In general, solar cookers can be classified into three types, that are box type, concentrating type and indirect solar cooker. Although many of scientists classified the solar cookers but R. M. Muthusivagami et al. has did the same in details as shown in the figure 1(Thirugnanam et al., 2020).

Figure 1. Types of solar cooker

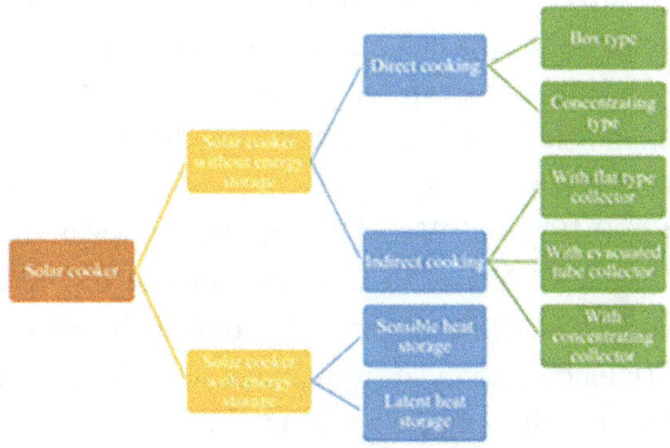

An angled sidewall box-type solar cooker equipped with reflective aluminium sheets was developed and optimized for Surat, India. The sidewall angles were fixed at 67.3° (south), 22.69° (north), and 35.44° (east & west) to increase solar reflection and thermal performance. The design of the cooker was meant to obtain higher temperatures and enhanced cooking efficiency compared to a traditional box-type solar cooker. The research assesses its thermal efficiency, figures of merit (F_1 & F_2), and cooking performance under various conditions. The designed optimum shows greater heat retention and efficient utilization of solar energy, and hence it has great potential to be used in solar cooking applications(Taşkesen et al., 2023). This study experimentally investigates the impact of different reflectors on the performance of a box-type solar cooker. The cooker was tested under three conditions: no reflector, aluminium foil reflectors, and mirror glass reflectors. Load and no-load tests were conducted using water as the absorbing medium to evaluate thermal performance. This experimental study investigates the impact of different reflectors on the performance of a box-type solar cooker. The results demonstrated that the addition of a mirror glass reflector significantly enhanced cooking efficiency. Specifically, the cooker was able to boil 1.5 kg of water in 51 minutes, achieving a second figure of merit (F_2) of 0.533. Compared to the no-reflector setup—which failed to boil water and proved unsuitable for cooking—the mirror glass reflector increased efficiency by 134%. The findings confirm that a mirror glass reflector-equipped cooker can effectively prepare various dishes with improved performance (Wassie et al., 2022). B.G. Venkateshwaran et al. studied the solar-based flat plate cooking unit (FPCU) with hook turbulators to enhance heat transfer. CFD analysis and experimental validation (cooking dosa) indicated that the use of turbulators enhanced heat transfer by 6.5%, decreased cooking time by 14%, and enhanced efficiency (6.2% heat transfer, 2.16% cooking unit, 1.55% overall). Economic analysis demonstrated a 13.38% lower cost per meal and 5.8% decreased cooking energy cost. The system also cuts CO_2 emissions by 503.22 kg/year (compared to LPG) and 420.17 kg/year (compared to electric stoves), making it an efficient and environmentally friendly cooking solution(Venkateshwaran et al., 2025). The present investigation compares the thermal performance of a box-type solar cooker (BTSC) using and without the use of Fresnel lens magnifiers (FLMG). There were three experiments: stagnation, load, and cooking power. When using FLMG, the first and second numbers of merit enhanced from 0.11 to 0.12°C·m²/W and 0.43 to 0.45, respectively Cooking power rose from 43.83 W to 46.87 W at a 50°C temperature difference, with an average magnifier transmittance of 72.26% and a 48.7% boost in the geometrical concentration ratio. Energy efficiency improved from 29.6% to 32.04%, confirming FLMG's effectiveness in enhancing thermal performance(Engoor et al., 2025).

However, the major limitation of solar cookers is their dependency on sunlight, which is always not available. This challenge can be effectively addressed by incorporating thermal energy storage (TES) systems, however there are many types of TES systems are there shown in figure 2. By storing excess thermal energy during peak sunlight hours and releasing it later, these systems ensure continuous cooking, even in the absence of direct sunlight. This advancement enhances the efficiency and feasibility of solar cookers, making them a reliable and sustainable cooking solution for the future(Sharma et al., 2000). With the development of thermal energy storage (TES) and solar collector technologies, solar cooking has increasingly become an alternative mode that is more feasible and cost-effective. The medium-temperature thermal energy (80–250°C) can be generated efficiently by the solar collectors and stored in TES systems in the temperature range of 150 to 200°C. This stored energy ensures consistent and reliable cooking even without solar radiation. As these technologies advance further, solar cooking is a promising and eco-friendly alternative to conventional cooking method(Zhou et al., 2023). Sharma et al. proposed a cylindrical PCM storage container for a box-type solar cooker (Fig. 4c), in which the PCM is encapsulated around the cooking pot, improving heat transfer and shortening the cooking time. The container is made of two concentric aluminium cylinders containing 2 kg of acetamide (melting point: 79°C, latent heat: 230 kJ/kg) and having fins on the inner wall for better heat transfer. The highest attained temperature was 127°C. Buddhi and Sharma subsequently created a latent heat storage unit for a three-reflector box-type solar cooker with the use of acetanilide (melting point: 118°C, latent heat: 222 kJ/kg) as the PCM. The reflectors, which can rotate along horizontal and vertical axes, improved the absorption of solar radiation, with a maximum temperature of 105°C(Sharma et al., 2000).Latent heat storage (LHS) systems leverage advanced heat exchangers and composite PCMs to enhance heat transfer. This study employs a copper tube–aluminium fin heat exchanger in a rectangular LHS unit, using HYTHERM 600 as the heat transfer fluid (HTF) at 80°C and 90°C. Heat transfer during melting transitions from conduction to natural convection, while solidification follows a conduction-dominated process. Among five stearic acid–paraffin wax (SPC) composites, S3 (50% stearic acid + 50% paraffin wax) demonstrated optimal performance, with the shortest melting time (185 min), longest solidification time (314 min), highest efficiencies (72.51% melting, 85.56% solidification), and maximum energy storage/rejection capacities (293.74/257.03 kJ/kg) at 90°C HTF. Its superior performance at higher HTF temperatures makes S3 a strong candidate for solar LHS applications in space and water heating(Saini et al., 2025).

2. BASICS OF THERMAL ENERGY STORAGE (TES)

With the increasing global requirement for renewable energy, scientists are obviously working hard at either identifying new energy sources or storage ways – both are equally critical, but appropriate energy storage is important because it not only fills the supply-demand gap but also benefits operating efficiency and reliability of systems and conserved energy. It also helps the system operator avoid costly fuels, waste energy and reduce costs. For example, energy storage can improve the operating efficiency of the generation of energy by using load levelling and develop energy value better managed economically. To be effectively employed as latent heat stores, Phase Change Materials (PCMs) must fulfil specific thermodynamic, kinetic, and chemical criteria; cost and availability are also an important factor when selecting PCMs. The desirable properties of PCMs can be summarized as follows(Sharma et al., 2009; Thirugnanam et al., 2020).

- Thermal Properties Appropriate phase-transition temperature consistent with operating conditions
- High latent heat of transition for maximum energy storage good heat transfer properties for charging and discharging Efficient heat-exchange rates with high thermal conductivity.
- Favourable phase equilibrium for consistent performance.
- High density for compact latent heat storage Minimal volume change during phase transitions to avoid mechanical stress Low vapor pressure to avoid storage containment issues.
- Kinetic Properties Minimal/no supercooling (for PCM that has too much supercooling, heat cannot be extracted) Sufficient crystallization rates to allow reliable phase transitions.
- Chemical Properties Good long-term chemical stability to avoid chemical degradation material Compatibility with construction materials to avoid re-

actions or corrosion non-toxicity for safe handling and environmental impact non-flammability for fire safety.

- Economic Factors
- Abundant and available for a scale of use affordable so we can afford to use it.

Balancing whether a PCM has the desired properties above to obtain performance, efficiency, and safety in a thermal storage application is key.

Figure 2. Types of thermal energy storage

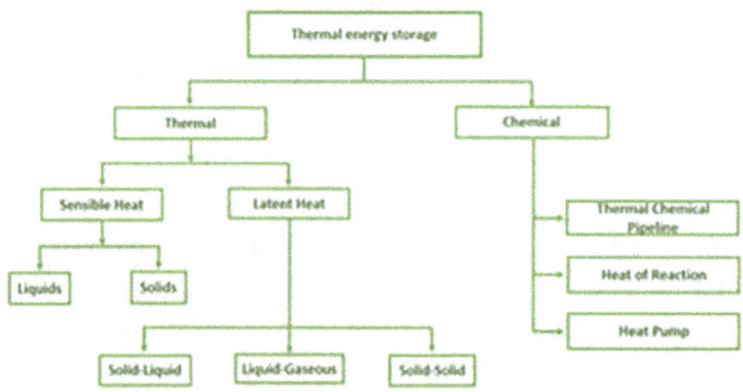

To this end, the storing of thermal energy using phase change materials (PCMs) can be a promising way of utilization of energy. However, before PCMs can be utilized in environments widely, there are numerous roadblocks from research & development of the PCM and they're not just use of performance but suitability in integration as a useful energy system at larger scale.

Phase Change Materials (PCMs) are materials that release and absorb thermal energy through phase changes, normally between solid and liquid phases. They absorb heat (latent heat of fusion) when melting and release stored energy when solidifying, making it possible to store thermal energy efficiently.

PCMs are categorized into three broad categories: organic, inorganic, and eutectic (figure 3). Organic PCMs, like paraffin wax and fatty acids, are chemically stable, non-corrosive, and thermally stable. Inorganic PCMs, like salt hydrates and metallic alloys, have greater thermal conductivity and energy density but are corrosive and susceptible to phase separation. Eutectic PCMs, which are blends of organic and/or inorganic compounds, are formulated to possess certain melting points and maximize thermal characteristics.

Figure 3. Types of the PCM

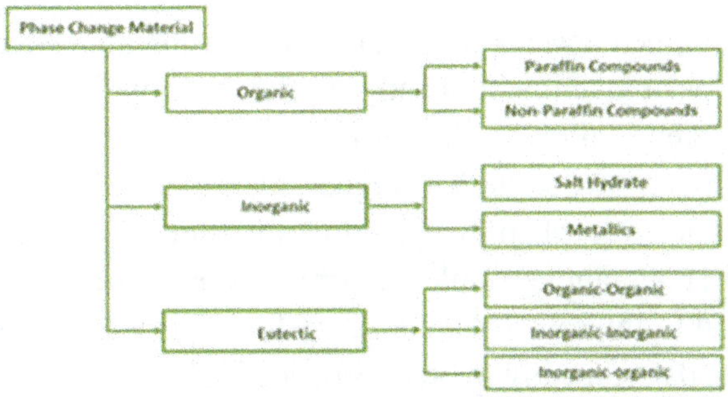

The operation of PCMs is based on a charging and discharging cycle. On heating, the PCM takes in energy and melts, retaining latent heat, and on cooling, it solidifies and gives out the stored heat. This feature of sustaining near-constant temperatures during phase changes makes PCMs suitable for different thermal management applications. PCMs have extensive applications in building energy efficiency for wall, roof, and floor thermal control, solar energy storage in solar water heaters and cookers, and thermal energy storage (TES) systems for HVAC and industrial applications. They also find a key application in the cold chain and medical storage, where they maintain stable temperatures for temperature-sensitive products such as vaccines. By increasing thermal energy storage and decreasing dependence on traditional heating and cooling systems, PCMs offer a green solution for enhancing energy efficiency in different applications.

3. ROLE OF TES IN SOLAR COOKERS

Food is a basic need. Traditional cooking methods use standard fuels, which have implications for fuel use and the environment. Solar cookers are an environmentally friendly alternative, harnessing solar radiation to cook food. However, they are at a disadvantage at night, or when cloudy without any solar radiation. Moreover, during peak solar radiation periods, solar cookers lose heat energy because they lack appropriate thermal storage. To mitigate these limitations, including thermal energy storage (TES) systems in solar cookers can assist. Of the various TES strategies, Phase Change Materials (PCMs) are of interest as PCMs can fit well into the concept of providing useful latent heat in a varied period. PCMs can enable solar cookers to utilize thermal energy to cook food at night, or during periods when solar radiation is limited, and make the energy-efficient solar cooker a viable solution for cooking food. Some of recent work is discussed here

Cuce et al. developed novel work that is distinct as he integrated Bayburt stone as a thermal energy storage medium and significantly improves the operation of a box-type solar cooker to allow for longer periods of cooking and stored heat. The cooker containing Bayburt stone had higher energy (35.3%–21.7%) and exergy (21.2%–14.1%) efficiency when compared to a traditional cooker, demonstrating its effectiveness in maintaining stable temperatures. Bayburt stone is promising as a versatile material to improve solar cookers, allowing them to operate more efficiently, reliably, and sustainably in practice(Cuce, 2018).

The research indicated that the choice of thermal energy storage (TES) material has an important role in the efficiency of solar cooking pots. The pot with sunflower oil had improved storage efficiencies at higher temperatures (3.2–4.4%) than the erythritol pot (2.5–3.9%) when using water for the cooking load. The heat utilization efficiencies with water (erythritol: 13–49%, sunflower oil: 17–46%) were higher than with sunflower oil at all periods (erythritol: 9.2–19%, sunflower oil: 9.4–28%) which also highlights the difference between the TES materials. The erythritol pot was more efficient when using water, whereas the sunflower oil pot was more efficient with sunflower oil. These results suggest that cooking medium and TES material selection can improve solar cooker performance. The use of TES material allows for prolonged cooking beyond direct solar exposure to provide practical and thermally efficient options for sustainable cooking(Mawire et al., 2022). Sathish's use of MWCNT/SiO_2 hybrid nano-PCM has a strong positive effect on the thermal performance of a box-type solar cooker. The best formulation of nano-PCM was able to reach a maximum PCM temperature of 128.9°C, increase cooking power up to 47.6 W, the efficiency reached 28.5%, and an energy efficiency of 6.2%. In addition, the cooking time was reduced to almost half, from 36.3 min to 18 min, confirming its ultrafast heating capability. Overall, the improvement achieved in performance demonstrates the potential of nano-enhanced PCMs to improve heat storage and efficiency and ultimately make solar cooking more practical and sustainable (Sathish, 2025).

C.B. Lwiwa and O.J. Nydal et. al. illustrates the efficiency of a reduced 3D Compound Parabolic Concentrator (CPC), which incorporates eight flat mirrors, for solar cooking. A 3D CPC is a type of solar concentrator designed to capture and focus sunlight onto an absorber. Unlike standard parabolic reflectors, which concentrate sunlight in one direction, a 3D CPC works in two dimensions, allowing it to collect and redirect sunlight from a wider range of angles without the need for continuous tracking by simplifying the CPC design with eight flat mirrors, an optimized reflector arrangement was achieved, eliminating the need for solar tracking. The system effectively collects and concentrates the sun's energy onto an aluminium cylindrical absorber, converting it to heat up to 135°C under 500 - 700 W/m^2 of solar radiation. Earlier cooking tests with a similar system have shown that even 140°C is sufficient for boiling water, and equatorial regions will likely achieve higher efficiency. The idea of heat storage and insulation allows for daytime cooking to continue in the evening after the sun has set, strengthening the CPC's potential as a low-cost and low-maintenance solar cooking solution(Garg et al., 1985).

4. DESIGN CONSIDERATIONS FOR TES INTEGRATION

Although many of cooker is developed by the researchers across the globe but they can be categorized as shown in figure 1. Some of the cooker's design are discussed below

4.1. Solar Cooker Without Extra Heat Storage

4.1.1 Panel Cookers

The panel cooker is direct solar cooker shown in figure 4. P.K. Gupta, A. Misal and S. Agrawal et. al. developed two design of panel cooker. This study is constructed from lightweight and durable corrugated polypropylene (PP) sheets, which enhance structural strength. A reflective sheet is glued to the PP base using food-grade adhesive to improve solar radiation reflection. The cooking pots, made of aluminium due to its lightweight and high thermal conductivity, are painted with epoxy black to enhance heat absorption. To create a greenhouse effect, a glass box is placed around the cooking area, helping to trap and retain heat. Two different panel cooker designs were developed and tested. Design A as show in figure 5 features a simple cubic structure with three open faces and a 1 sq. ft. reflecting area, serving as a basic reference for performance comparisons. Design B (figure 6), in contrast, incorporates a folded circular quadrant, optimizing sunlight reflection toward the central cooking area. This modification provides better efficiency in both summer and winter solstices by improving the angle of solar reflection. Although these designs effectively harness solar energy, they do not incorporate additional thermal energy storage (TES) systems such as phase change materials (PCM) or sensible heat storage. Instead, they rely entirely on direct solar heating, with Design B demonstrating superior heat retention and efficiency due to its improved reflective geometry(Gupta et al., 2021).

Figure 4. A schematic view of solar panel cooker

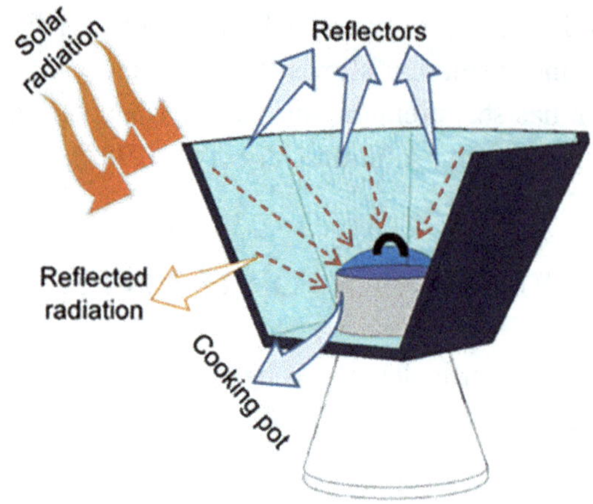

Figure 5. (a) section of design A and (b) Complete Unit of Design A.

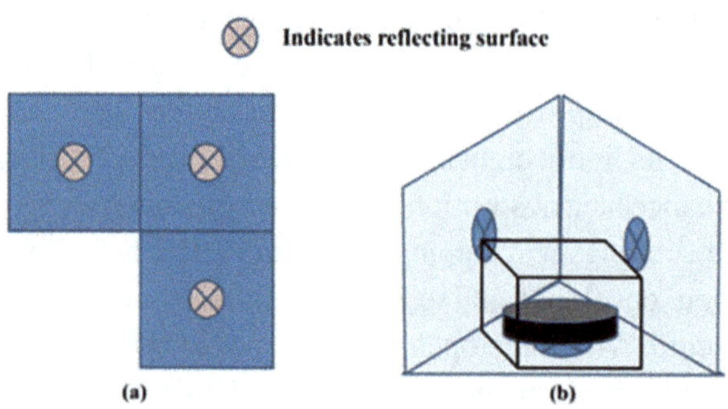

Figure 6. (a) Development and (b) Complete unit of design B panel cooker.

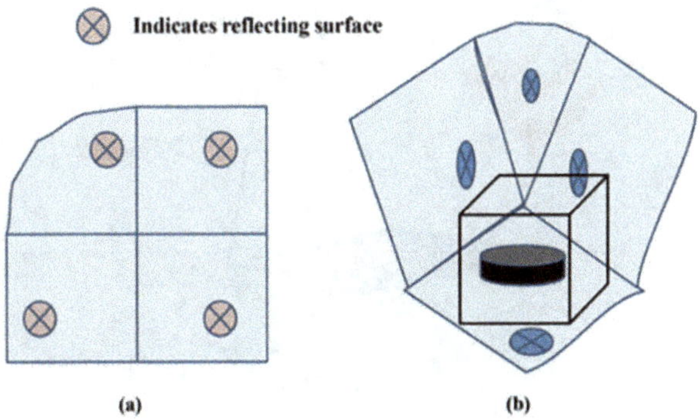

4.1.2. Box Type Cookers

A box-type solar cooker shown in figure 7 is a straightforward solar cooker that captures and uses the sun's energy to cook food by holding heat in an insulated box. It consists of a blackened cooking chamber, a transparent glass or plastic lid, and reflectors that intensify solar energy. When sunlight passes through the transparent glass cover, the black interior retains it, and this transforms visible light to heat. The heat that is retained inside becomes a greenhouse gas effect that limits heat loss and thus raises the temperature inside. With resistant insulation, the solar cooker can retain the heat effectively and cook food by several hours of slow cooking processes from boiling, baking, to steaming.

Figure 7. Box type solar cooker

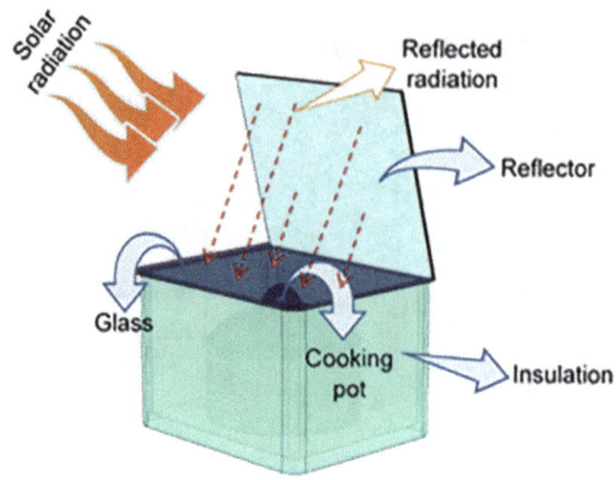

S. Geddam et al. desgined double exposure box-type solar cooker consisting of a double-glass fitting of 0.1344 m^2 for solar radiation to reach the absorber. The absorber is a black-painted galvanized sheet of steel while the cooking vessel is a cylindrical pot made of aluminum with a flat bottom with a diameter of 0.18 m and 0.06 m in height. In an effort to enhance efficiency and shorten cooking times, the cooking vessel was compared with a finned cooking vessel. The modified cooking vessel was identical to the original pot used but consisted of 36 aluminum fins of rectangular cross-section (5.5 cm × 0.05 cm) painted black (top and outside) and attached to the surface reaching a total length of 2.2 cm. The fins created a small amount of increased surface area and were also intended to help with heat absorption. The theory was that the increased surface area would provide the cooking vessel with improved thermal performance and require less time to cook food than the standard cooking vessel, under the same solar conditions(Geddam et al., 2015).

The solar cooker is constructed according to the Goswami et al. and Vaidya et al. protocols. It has a 340 mm × 340 mm × 70 mm metal absorber tray, black matte paint coated, and covered with a 560 mm × 580 mm top structure. It is insulated with Cerawool (64 kg/m^3, 0.48 W/m-K thermal conductivity)—50 mm thick below the absorber and 25 mm on the sides—to reduce heat loss. The housing is Aluminum Composite Panel (500 mm × 500 mm) and double-glazed float glass (4 mm thick, 15 mm spacing) is utilized as the cover, sealed with adhesive in a grooved frame. Calibrated T-type thermocouples are utilized for temperature measurement at different points. The new cooker is of identical design, with optimized sidewall angles and surface finishing by polishing, buffing, and chemical cleaning. The glass frame is hinged to the housing for simple opening(Taşkesen et al., 2023).

Avnish Kumar et al. aimed to enhance the heat transfer rate inside the cooker by combining extended surface geometry with latent heat storage. Notably, no such technique has yet been reported for solar box cookers specifically designed for hilly regions.

The tested solar hot box cooker (SHBC) was fabricated using locally available materials, following the conventional design standards proposed by Garg et al. (1985). The cooker featured a darkened interior with an aperture area of approximately 36 cm^2 and double-glazing made of transparent glass (50×48 cm^2). The aluminum cooking plate was placed 15 cm below the glazing sheet, and a mirror reflector (51×53 cm^2) was mounted on the lid for additional solar gain. The cooker had a total height of 20 cm and a specific area of 66 cm^2. Glass wool was used as insulation to minimize thermal losses. The cooker accommodated four identical aluminum vessels (1 L each), coated dull black, with a total cooking capacity of 3–4 kg.

The PCM bed, the innovation of the research, consisted of 140 small copper capsules (9.6 mm $\times$ 30 mm) filled with mostly (92%) commercial-grade paraffin wax RT64 as the PCM, which has a melting point of 64 °C, thermal storage capacity of 251 kJ/kg, specific heat capacity of 2.1 kJ/kg·K, thermal conductivity value of 0.2 W/m·K and density that varied from 0.88 kg/L (solid at 20 °C) to 1 kg/L (liquid at 88 °C). The total mass of PCM used was approximately: 1.5 kg. The 140 capsules were held onto a thin (0.5 mm), black-coated aluminium sheet and although the copper capsules were attached via thermally conductive glue, they were mounted loosely to make disassembly and servicing easier.

In conclusion, once we began preparing experimental cooking trials with the modified cooker, we measured one maximum cooking plate temperature of 150 °C, an instantaneous thermal efficiency of the system of 45.7%, cooking power of the system was 54.71 W, heat transfer coefficient (HTC) was 311 $W/m^2 \cdot °C$; overall heat loss coefficient (U_loss) was 5.71 $W/m^2 \cdot °C$; the overall cost of the modified cooker was approximately \$48.19, with a payback time of approximately 3 years and 11 months; and the modified cooker was able to adequately prepare a variety of common dishes from the region; and demonstrated effectiveness for household use(Kumar et al., 2023).

Conventional box-type solar cookers are limited by inherently only allowing for activities during the day, reducing their functionality and interest to consumers. New direction utilizing waste pieces of marble, as sensible heat storage media, was proposed to remedy this. These pieces of waste marble can be employed as integrated storage, or portable storage, improving the thermal effectiveness and evening cooking function.Additionally, a matt black-coated aluminum sheet was installed below the glass cover to reduce top heat losses.

Two configurations were tested:

(I) Cooking load with heat storage (no aluminum sheet), and
(II) Cooking load with both aluminum sheet and heat storage.

The solar cooker used in the study had an internal chamber size of 350 × 350 × 100 mm and external dimensions of 520 × 520 × 190 mm. The glass cover measured 440 × 440 × 5 mm, and the aluminum sheet placed below it was 500 × 500 × 0.4 mm. Two cooking pans (radius: 140 mm; height: 70 mm) were used side by side. For heat storage, 4.33 kg of waste marble pieces (density: 2711 kg/m^3) were placed inside the cooker. For simplification in numerical modeling using COMSOL Multiphysics 6.0, the marble pieces were considered as a perfect cuboid.

Experimental results in outdoor conditions revealed that configuration I achieved a cooking time of 154–155 minutes, cooking power of 56.62 W, and thermal efficiency of 25.79%.Although Configuration II that included the aluminum sheet had slightly longer cooking times (175–176 minutes) and lower cooking power (48.21 W), it had a higher thermal efficiency of 31.19%. This suggests that although the aluminum sheet did slightly reduce the cooking power due to its reflective properties, it did help retain heat, which increased thermal efficiency.

The use of waste marble as a portable heat source also proved beneficial in post-cooking thermal availability. Once the first meal was cooked and removed, the stored heat in the marble pieces maintained high internal temperatures until the evening, making second or delayed cooking possible. The marble pieces, when used independently after initial heating, provided enough residual heat to support evening cooking or preheating of food, which could then be finished using an electrical heating coil if needed(Kumar Goyal & Eswaramoorthy, 2024).

4.1.3. Concentrating Type Cookers

A Concentrating Solar Cooker (CSC) operates by utilizing a reflective mirror, such as a parabolic dish, Fresnel lens, or a multi-faceted reflector, to concentrate the sun's light and direct that into a cooking vessel. These devices deliver cooking temperatures of up to 300°C with a concentration ratio of up to 50. The basic parts of a CSC are the following (Figure 8):

- Reflecting Structure: a parabolic dish to reflect and concentrate sunlight.
- Cooking Vessel: placed at the focal point of the concentrating device and properly designed to efficiently absorb the concentrated heat energy.
- Rigid Supporting Structure: maintains the device in a stable position.
- Coaster Wheels: allows for effective solar tracking for efficient cooking. For cooking for lots of people, the device is an indirect system that would include.

Figure 8. Diagram of concentrating solar cooker

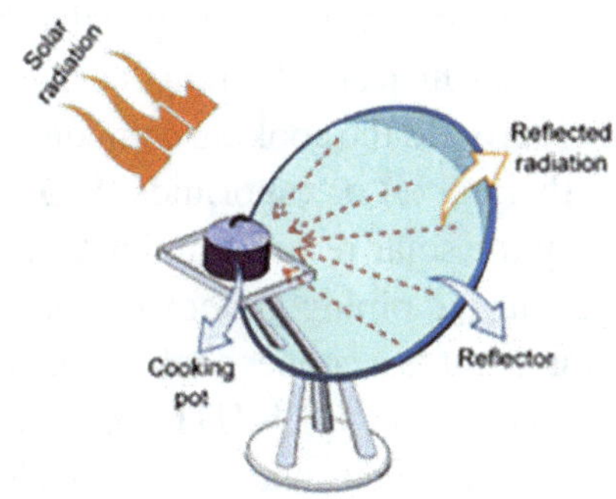

A large Concentarting Solar Cooker (CSC) features an indirect (indoor) unit to realize a continuous and efficient cooking experience. The system consists of

- High-Temperature Solar Collector
- Heat Transport Unit
- Thermal Energy Storage (TES) –
- Heat Transfer Unit
- Heat Transfer Fluid (HTF)

Special Features for Large-Scale Cooking:Capacity for 50 persons, suitable for commercial/full kitchens, hostels, hotels, pilgrimage centers, orphanages, and industries.Designed for sustainable mass cooking with increased efficiency.Provides a cost-effective alternative to conventional cooking fuels(Saxena et al., 2011).

The initial documented endeavors of solar cooking originate back to Horace de Saussure's experiments in 1767. He created a box with which fruits were boiled utilizing solar radiation, and achieved a temperature of about 88 oC.

The Parabolic Solar Trough Cooker (PSTC) is designed by N. El Moussaoui, et al. in which a 200 cm long, 35 kg system designed to reach over 200°C with heat storage for stable cooking. It features a 1.3 m^2 parabolic concentrator that reflects sunlight onto a 150 cm vacuum tube (47 mm inner, 58 mm outer diameter), where a thermal oil circulates.The cooking pot (5L) is heated by the oil that enacts the heat transfer through copper tubes: one tube for the return flow of the cold fluid that, in turn, drives two other tubes carrying the hot fluid. The insulated wooden cooking box limits heat loss, and the orientation wheels make it easy to move the pot to achieve optimal solar tracking(El Moussaoui et al., 2020).

M. Hosseinzadeh et al. designed an indirect solar cooker functions by concentrating solar radiation with a parabolic dish concentrator (120 cm aperture, 45 cm focal length). Solar radiation is projected onto an elliptically shaped receiver (400 ml volume) which has spiral blades, which are (1.5 mm thick, 12 mm high), to increase heat transfer from the solar receiver to the cooking component. A heat transfer fluid (HTF), consisting of either thermal oil or nanofluids (SiO_2-oil, TiO_2-oil, SiC-oil), absorbs the heat generated by the solar receiver and is then pumped as a constant flow of 250 ml/min throughout the piping system.The HTF then passes through copper tubing from the cooking unit to a cooper cooking vessel wrapped by a heated helical copper coil (6 mm inner diameter). This HTF continuously flows through the copper coil and transfers the heat to the insulated cooking chamber (K-Flex insulation) of the cooking unit so that the food can extract heat efficiently.Temperature is monitored by using K-type thermocouples at different measuring points throughout the system (solar receiver, cooking vessel, and HTF). Additionally, a pressure transmitter, mercury thermometer, and solar power meter take measurements to determine system performance. Experiments were conducted at the Research Institute of Food Science and Technology in Mashhad, Iran (Latitude: 36°, Longitude: 59°), where it was tested to boil 2 L of water in the cooker. The nanofluids were prepared using high-speed stirring and ultrasonication (4 cycles of 30 min at 50°C) with CTAB as a surfactant to create stability without sedimentation(Hosseinzadeh et al., 2021).

The experimental system designed for an indoor cooking setup with a parabolic dish collector (PDC) includes a solar collector, cooking pot, and energy exchange system. It is based in the indoor kitchen of NIT Kurukshetra (29.96°N, 76.87°E) and allows heat to be transferred from a collector located outdoors to an indoor cooking appliance with the ability to use a heat transfer fluid (HTF). In this arrangement, the solar collector constitutes a parabolic dish made of FRP (2.6 m^2), with 0.583 m focal length, and a depth of 0.35 m, which is covered with 10 cm × 10 cm sheets of solar-grade aluminum reflector sheets (95% reflectivity). The receiver is a conical copper tube (8 mm diameter, 200 mm height, 200 mm diameter) that is insulated with 40 mm glass wool and painted black for maximum absorption. The PDC is stabilized with a mild steel stand containing 12 mm stainless steel fasteners and is adjustable for manual tracking with a gearbox with a 40:1 ratio. The cooking pot has a volume of 1.5 L, two walls, is made of food to the cooking pot grade stainless steel (0.9 mm thickness), and contains an oil jacket that circulates the HTF. The cooking pot is also insulated with glass wool and has a stainless steel lid to minimize heat loss. The HTF will circulate through the collector, pump, cooking pot, allowing for indoor cooking with sufficient heat transfer performance to achieve a continuous cooking process

5. ADVANTAGES OF TES FOR OFF-SUNSHINE COOKING

The use of Phase Change Materials (PCMs) in Thermal Energy Storage (TES) in solar cooking allows for off-sunshine cooking by effectively capturing solar energy for use when there is no direct sunlight. A.G. Bhave and C.K. Kale et al. designed thermal storage type solar cooker for high temperature cooking using solar salt. The system was able to maintain high cooking temperatures, above 170–180 °C for frying, so that cooking can occur indoors at a more convenient time. Based on experimental trials, 250g of potato chips were fried in about 17 minutes, and rice was cooked in two consecutive batches of 20 minutes each. Furthermore, the TES system enhances cooking efficiency by ensuring continuous heat transfer from stored energy in the PCM to the cookware. The efficiency of collecting and storing solar energy is currently between 11% and 14% and performance will improve with further work on insulation and concentrator fabrication. (Bhave & Kale, 2020). R. Senthil et al used Phase Change Material (PCM) helps to increase the performance of a Parabolic Dish Solar Cooker (PDSC), particularly in off-sunshine cooking. This system consists of a parabolic collector with a 0.9 m diameter, and a cooking vessel with fins, which utilizes paraffin wax as PCM to store and release thermal energy effectively. The solar cooker demonstrated an average energy performance of 22% and exergy efficiency of 2.6%. Water heating studies show that the solution without PCM reached a temperature of 90°C in 90 minutes, while PCM usage improved heating time to 120 minutes – this shows the potential of thermal storage in improved heating performance. Off-sunshine cooking performance improved by a factor of 2 with PCM. A total heat requirement for cooking 1.6 kg of rice was outlined, indicating that cooking 1 kg requires about 272 W of thermal energy. A combined convective heat transfer assumption was used (forced air and buoyancy-driven convection), with a heat transfer coefficient between 8-10 W/m^2K. Experimental assessments done at [12° 800 N, 80° 020 E, 6 m above sea level] using in-field conditions indicated the solar beam radiation intensity was between 350-780 W/m^2, wind speed was between 0.5-2.1 m/s, and ambient temperature was between 33-36°C. The effect of fins in the cooking vessel enhanced the heat transfer process significantly, and the time taken to reach 90°C was 30 minutes shorter than with a vessel without fins. The monitoring of the heat release also showed that the cooling rates were dependant on the quality of the insulation, ambient temperature, and vessel loading. From the observations with the charging and discharging with the PCM, it was noted that, while the rate of heat transfer was initially high, the rate of heat transfer decreased for the first hour which ultimately confirmed that the PCM is effective at storing heat for longer-term cooking. Stored thermal energy can also be repurposed for various cooking applications, including baking flatbreads (like chapatis) or using an electric pressure cooker indoors. Beyond household cooking,

TES could be extended to thermal applications in industry, such as steam generation. The TES systems can store excess thermal energy during peak sunlight hours to meet the nighttime energy needs, which will improve overall efficiency and expand the functional capability of solar thermal technology.

The effectiveness of Thermal Energy Storage (TES) to cook off-sunshine was assessed using temperature monitoring of heat lost from the solar cooker. The cooling rate was influenced not only by the degree of vessel insulation, but also by the ambient temperature and the loading of each vessel. Both the heating and cooling data collected over the timespan of the PCM charging and discharging trials indicated a rapid heat transfer rate at the beginning of the study, which then declined over the first hour. This data supported the ability of phase change materials to store thermal energy over time. Thermal energy stored in PCM is useful in various home cooking applications such as baking flatbreads (i.e. chapatis), heating an electric pressure cooker, and so on. TES has more applications, beyond home cooking, such as harnessing heat generation in industrial applications, including steaming food in restaurants, to hold thermal energy produced as surplus, and providing heating energy during off-sunshine hours, such as overnight, improving the efficiency of the heating demand for the specific time slot and providing flexibility of operation in solar thermal technologies.

A high performance concentrating solar cooker prototype was developed from specifications provided to Heliac (2020). The entire system consists of a wooden reticular support that is equipped with wheels for movement, with two wooden frames, one holding a special rectangular Fresnel-type lens, and a second smaller frame holding a mirror with a reflective surface. The lens concentrates solar radiation onto the mirror's focal area, which then reflects the rays onto a designated cooking surface where a pot is placed. The working mechanism is illustrated in Figures 9.

Figure 9. Heliac solar cooker

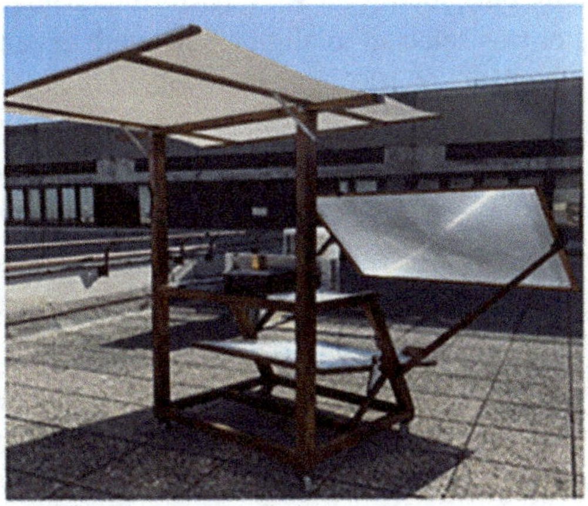

The lens, made from a polymeric film mimicking the shape of a traditional Fresnel lens, was fabricated through Heliac's low-cost optical film production process. Its dimensions are 1120 × 1390 mm, forming the aperture area (Aa) of the cooker. The geometrical concentration ratio (C) of the system was calculated using:

$$C = \frac{A_a}{A_p} = \frac{1.557}{0.038} = 40.97$$

where Ap is the area of the lower surface of the pot exposed to concentrated solar radiation. This high concentration ratio classifies the cooker as a high-performance solar concentrator.

The mirror, having dimensions of 730 × 900 mm, is made from a unique reflective film that is capable of reflecting both infrared and ultraviolet radiation. The mirror surface needs to be kept clean and under tension to achieve maximum reflection efficiency. The lens is mounted on the wood structure with two wooden arms (1408 mm long) which allow rotational adjustments from 0° to 90°, to achieve the optimal solar position. Manual adjustments every 5–10 minutes will maintain the best solar follow mechanism and performance.

Outdoor testing of the prototype was conducted on the roof of the DIISM facility (Latitude: 43.5867° N, Longitude: 13.5150° E) during June–July 2020 and March–April 2021. Throughout the tests, direct solar radiation, ambient temperature, and test fluid temperature were continuously recorded. Measurements were performed using thermocouples and a pyrheliometer, with data acquisition handled by a Pico Technology TC-08 data logger connected to a laptop. The cooking vessel used was a 22 cm diameter stainless steel pot coated with black, heat-resistant paint based on silicone resin, capable of withstanding up to 600°C. Since the cooker is a concentrating system, only load tests were performed using water and silicone oil as test fluids. Water was selected to align results with existing literature, while silicone oil (Rhodorsil Oil 47 V 100) enabled evaluation at temperatures beyond 100°C.Experiments showed that it could boil 3 kg of water in 30 minutes and heat 3 kg of silicone oil from 40°C to 170°C in under one hour. The ability of the cooker to obtain strong thermal efficiency and high working temperatures confirms that the lens-mirror system works effectively. Furthermore, the cooker had a high opto-thermal ratio and thermal efficiency, comparable to other concentrating systems reported in the literature.

This prototype offers several practical advantages: it is cost-effective, constructed from recycled materials, compact, and easy to manufacture and operate. These features make it particularly suitable for use in developing regions, where affordability, simplicity, and sustainability are critical. The experimental findings, including detailed thermodynamic performance data, can serve as a valuable reference for further development and comparative analysis in the field of solar cooking technologies(Coccia et al., 2021).

6. CHALLENGES AND LIMITATIONS

Although SCs are the effective but it also have some challenges and Limitations tabulated in table 1 (Aramesh et al., 2019; Odoi-Yorke et al., 2025).

Table 1. Challenges and Limitations in Solar Cookers with PCM Integration

Category	Specific Challenge/ Limitation	Key Issues
Material Development	PCM performance improvements	Low thermal stability, suboptimal latent heat storage, non-ideal melting points
	Eco-friendly alternatives	High cost of bio-based PCMs, limited thermal performance
Computational Gaps	Predictive modelling	Lack of accurate simulations for solar cooker performance
	Experimental methods	Reliance on costly trial-and-error approaches
Technological Integration	Hybrid systems (PV + TES)	Complex integration, low efficiency in combined systems
	Smart technologies	High cost of automated tracking/IoT systems
Socio-Economic Barriers	Regional adoption	Limited research collaboration in developing nations

continued on following page

Table 1. Continued

Category	Specific Challenge/ Limitation	Key Issues
	Cultural acceptance	Mismatch with local cooking practices, economic constraints
Standardization Issues	Performance evaluation	No universal testing protocols
	Environmental variability	Inconsistent performance across climates
Funding Disparities	Research distribution	Unequal global funding, knowledge gaps in developing regions
Weather Dependency	Operational limitations	Reduced function in clouds/nighttime, PCM heat retention limits
Cooking Efficiency	Time and energy losses	Longer cooking times vs. conventional stoves, heat loss via conduction/convection
Affordability	High initial costs	Expensive hybrid systems/nano-PCMs limit accessibility
Design Constraints	Portability & versatility	Bulky designs, inability to cook diverse foods at required temperatures
Environmental Impact	Material sustainability	Synthetic PCMs/non-recyclable components conflict with green goals

7. FUTURE DIRECTIONS AND INNOVATIONS

Future work focusing on future solar cooking research should either improve design for heat storage or heat retention additionally improve thermal efficiencies. Future work will focus on the continued development of advanced thermal models for performance optimization and testing against experimental data. Notably, examining select storage energy storage materials (e.g. phase change materials) with differing thermal storage properties may assist with improving energy storage and distribution. Future issues may also be improved by the inclusion of new heat transfer enhancements, such as appropriate internal turbulators or optimized reflectors, which will improve solar cooking efficiency and reduce overall solar cooking times. For purposes of furthering portability, lightweight designs for throughout the full cooking process should be explored, while maximizing thermal performance in thermal form. Particularly with this regard, integration of PCMs for extended thermal retention in solar cookers, and hybrid solar cookers with auxiliary heating modes, should also be examined for improved sustainability and reliability in solar cooking, in line with the conditions of each site. Finally, the implementation of economic and environmental costs is now increasingly achievable, especially about accounting for cost savings, carbon footprints and environmentally improvements. This assessment is also essential for stimulating ownership - utilization of solar cooking technologies globally. Future research on PCESMs should focus on enhancing their thermal conductivity, stability, and encapsulation techniques to overcome existing limitations and expand their applicability. The development of hybrid and nano-enhanced PCESMs offers promising avenues for improving thermal performance and durability. Additionally, exploring bio-based PCMs as sustainable alternatives to synthetic materials can further align with environmental goals.

The integration of PCESMs into advanced energy systems remains a key area of interest, particularly in solar energy applications, building thermal management, and electronics cooling. Further studies should explore optimizing these materials for smart textiles, electric vehicles, and healthcare applications, such as temperature-sensitive drug delivery.

For solar cooking applications, future efforts should focus on optimizing reflector materials and designs to further enhance heat absorption and retention. The integration of advanced cooking appliances, such as pressure cookers, can improve cooking efficiency and broaden adoption in rural communities. Additionally, developing cost-effective manufacturing techniques for solar cookers with high thermal performance will be crucial for widespread implementation (Taşkesen et al., 2023; Mawire et al., 2022; Obayedullah, 2025).

8. CONCLUSION

Solar cooking technology presents a sustainable alternative to conventional cooking methods, offering significant environmental, economic, and health benefits by reducing reliance on fossil fuels, lowering carbon emissions, and improving air quality. Research highlights the comparative efficiency of different solar cooker designs, with box-type systems being simpler to implement while concentrator-based designs deliver higher cooking temperatures and faster performance. The integration of thermal energy storage (TES) systems, such as phase change materials (PCMs) and sensible heat storage media like sunflower oil, addresses the intermittency of solar energy, with sunflower oil enabling higher cooking temperatures and shorter cooking times, while PCMs extend heat availability beyond sunlight hours through efficient latent heat storage and enhanced thermal conductivity. However, widespread adoption faces challenges, including weather dependency, cultural preferences, and high initial costs, necessitating advancements in TES efficiency, optimized cooker designs, and expanded applications beyond household use to industrial scales. To accelerate adoption, a multifaceted approach is essential, combining technological innovations—such as cost-effective materials, localized design optimizations, and hybrid systems incorporating photovoltaics and batteries—with efforts to boost social acceptance through awareness campaigns, aesthetic improvements, and alignment with traditional cooking practices. Additionally, supportive policy measures, including subsidies, standardized testing protocols, and international research collaborations, are critical to reducing costs, ensuring reliability, and fostering global knowledge sharing. By addressing these technological, social, and policy dimensions, solar cooking can evolve into a practical, efficient, and widely embraced solution for sustainable cooking worldwide.

References

Aramesh, M., Ghalebani, M., Kasaeian, A., Zamani, H., Lorenzini, G., & Mahian, O.. (2019). A review of recent advances in solar cooking technology. *Renewable Energy*, *140*, 419–435. DOI: 10.1016/j.renene.2019.03.021

Bhave, A. G., & Kale, C. K. (2020). Development of a thermal storage type solar cooker for high temperature cooking using solar salt. *Solar Energy Materials and Solar Cells*, *208*, 110394. DOI: 10.1016/j.solmat.2020.110394

Coccia, G., Aquilanti, A., Tomassetti, S., Ishibashi, A., & Di Nicola, G. (2021). Design, manufacture and test of a low-cost solar cooker with high-performance light-concentrating lens. *Solar Energy*, *224*, 1028–1039. DOI: 10.1016/j.solener.2021.06.025

Cuce, P. M. (2018). Box type solar cookers with sensible thermal energy storage medium: A comparative experimental investigation and thermodynamic analysis. *Solar Energy*, *166*, 432–440. DOI: 10.1016/j.solener.2018.03.077

El Moussaoui, N., Talbi, S., Atmane, I., Kassmi, K., Schwarzer, K., & Chayeb, H.. (2020). Feasibility of a new design of a Parabolic Trough Solar Thermal Cooker (PSTC). *Solar Energy*, *201*, 866–871. DOI: 10.1016/j.solener.2020.03.079

Sathish. (2025). MWCNT SiO2 Hybrid Nano-PCM for Ultrafast Solar Cookers An Experimental Study (1).pdf n.d.

Engoor, G. G. S. S, and Veerappan AR. (2025). Experimental investigation of a box-type solar cooker incorporated with Fresnel lens magnifier. *Energy Sources. Part A, Recovery, Utilization, and Environmental Effects*, *47*, 496–511. DOI: 10.1080/15567036.2020.1826009

Garg, H. P., Mullick, S. C., & Bhargava, A. K. (1985). Sensible Heat Storage. *Sol Therm Energy Storage*, *7*, 82–153. DOI: 10.1007/978-94-009-5301-7_2

Geddam, S., Dinesh, G. K., & Sivasankar, T. (2015). Determination of thermal performance of a box type solar cooker. *Solar Energy*, *113*, 324–331. DOI: 10.1016/j.solener.2015.01.014

Gupta, P. K., Misal, A., & Agrawal, S. (2021). Development of low cost reflective panel solar cooker. *Materials Today: Proceedings*, *45*, 3010–3013. DOI: 10.1016/j.matpr.2020.12.004

Hosseinzadeh, M., Sadeghirad, R., Zamani, H., Kianifar, A., Mirzababaee, S. M., & Faezian, A. (2021). Experimental study of a nanofluid-based indirect solar cooker: Energy and exergy analyses. *Solar Energy Materials and Solar Cells, 221*, 110879. DOI: 10.1016/j.solmat.2020.110879

Kumar, A., Saxena, A., Pandey, S. D., & Gupta, A. (2023). Cooking performance assessment of a phase change material integrated hot box cooker. *Environmental Science and Pollution Research International, 31*, 62392–62407. DOI: 10.1007/s11356-023-25340-x PMID: 36689111

Kumar Goyal, R., & Eswaramoorthy, M. (2024). Theoretical and experimental analysis of box-type solar cooker with sensible heat storage. *Solar Energy, 268*, 112273. DOI: 10.1016/j.solener.2023.112273

Mawire, A., Lentswe, K., & Owusu, P. (2022). Performance of two solar cooking storage pots using parabolic dish solar concentrators during solar and storage cooking periods with different heating loads. *Results in Engineering, 13*, 100336. DOI: 10.1016/j.rineng.2022.100336

Obayedullah, M. (2025). Review Article Recent Advances in Phase Change Energy Storage Materials. *Developments and Applications, 2025*. Advance online publication. DOI: 10.1155/er/6668430

Odoi-Yorke, F., Baah, B., & Opoku, R. (2025). Scientific mapping of experimental research on solar cookers: Global trends, evolution, and future directions. *Sol Energy Adv, 5*, 100093. DOI: 10.1016/j.seja.2025.100093

Saini, D. K., Chandrashekara, M., & Yadav, A. (2025). Thermal analysis of a rectangular latent heat storage unit with stearic acid/paraffin wax composite PCM for solar thermal energy storage systems. *Energy Sources. Part A, Recovery, Utilization, and Environmental Effects, 47*, 4279–4296. DOI: 10.1080/15567036.2025.2460644

Saxena, A., Varun, Pandey SP, Srivastav G. (2011). A thermodynamic review on solar box type cookers. *Renewable & Sustainable Energy Reviews, 15*, 3301–3318. DOI: 10.1016/j.rser.2011.04.017

Sharma A, Tyagi V V, Chen CR, Buddhi D. Review on thermal energy storage with phase change materials and applications 2009;13:318–45. https://doi.org/.DOI: 10.1016/j.rser.2007.10.005

Sharma, S. D., Buddhi, D., Sawhney, R. L., & Sharma, A. (2000). Design, development and performance evaluation of a latent heat storage unit for evening cooking in a solar cooker. *Energy Conversion and Management, 41*, 1497–1508. DOI: 10.1016/S0196-8904(99)00193-4

Taşkesen, E., Dirik, M., Tekir, M., & Pazarlioğlu, H. K. (2023). Predicting heat transfer performance of Fe3O4-Cu/water hybrid nanofluid under constant magnetic field using ANN. *J Therm Eng*, *9*, 811–822. DOI: 10.18186/thermal.000000

Thirugnanam, C., Karthikeyan, S., & Kalaimurugan, K. (2020). Study of phase change materials and its application in solar cooker. *Materials Today: Proceedings*, *33*, 2890–2896. DOI: 10.1016/j.matpr.2020.02.780

Venkateshwaran, B. G., Kumaresan, G., Santosh, R., & Velraj, R. (2025). Performance enhancement of flat plate cooking unit using novel hook turbulators for indirect solar cooking applications: Numerical and experimental assessment with enviroeconomic analyses. *Solar Energy*, *287*, 113247. DOI: 10.1016/j.solener.2025.113247

Wassie, H. M., Getie, M. Z., Alem, M. S., Kotu, T. B., & Salehdress, Z. M. (2022). Experimental investigation of the effect of reflectors on thermal performance of box type solar cooker. *Heliyon*, *8*, e12324. DOI: 10.1016/j.heliyon.2022.e12324 PMID: 36582681

Zhou, C., Wang, Y., Li, J., Ma, X., Li, Q., & Yang, M.. (2023). Simulation and economic analysis of an innovative indoor solar cooking system with energy storage. *Solar Energy*, *263*, 111816. DOI: 10.1016/j.solener.2023.111816

Chapter 5
Study of the Thermal Performance of Different Solar Cooker Geometries With Single and Double Glass

ZMIM Fatima
https://orcid.org/0009-0006-0839-3197
Mohammed First University, Oujda, Morocco

Amraqui Samir
Mohammed First University, Oujda, Morocco

Mezrhab Ahmed
Mohammed First University, Oujda, Morocco

ABSTRACT

The growing use of fossil fuels has led to a significant rise in greenhouse gas emissions, directly contributing to global warming and environmental imbalances. The burning of coal, oil, and natural gas releases large amounts of carbon dioxide (CO_2), methane (CH_4), and nitrogen oxides (NO_X), intensifying the greenhouse effect and accelerating climate change. As a result, transitioning to renewable energy sources is crucial to mitigate global warming, protect public health, and preserve the environment. Solar cookers are an effective technology that captures and converts solar radiation into usable heat for various applications. This article presents a thermal study on three solar cooker designs: rectangular, trapezoidal, and triangular. The goal is to determine the temperatures of the absorber surface and indoor air. The

DOI: 10.4018/979-8-3373-2535-4.ch005

study compares the effects of single and double glazing on thermal performance and considers the impact of adding or removing a thermal load inside the cooker. Additionally, simulations with a primary reflector are also analyzed to assess its effect on performance.

1 INTRODUCTION

The use of fossil energy sources continues to increase every day around the world to meet our daily needs. This growing dependence is mainly due to their essential role in various sectors, including electricity production, transportation, and industry. Fossil fuels such as oil, coal, and natural gas are widely used to power our homes, operate vehicles, and fuel industrial machinery. However, this reliance poses serious challenges. The extraction and combustion of these resources significantly contribute to greenhouse gas emissions, leading to climate change and major environmental issues. Moreover, the gradual depletion of fossil fuel reserves raises concerns about the sustainability of our current way of life. It is therefore crucial to explore more sustainable and renewable alternatives to ensure responsible energy use that respects the environment and preserves resources for future generations.

This transition to renewable energy sources, such as solar, wind, and hydropower, is essential for reducing our carbon footprint and building a more sustainable energy future. Among the technologies that directly utilize renewable and sustainable energy sources, solar cookers stand out. These devices harness solar radiation and convert it into usable heat based on the greenhouse effect phenomenon. Solar cooking technology dates back to ancient times, demonstrating humanity's long-standing efforts to harness the sun's energy for practical applications, In 1774, Horace-Bénédict de Saussure studied the thermal effects of solar rays using his Helio thermometer, a device with insulating walls and one or more glass panels that allowed light to pass through, thus creating a greenhouse effect. This instrument could reach temperatures of about 80°C. His research laid the groundwork for the development of solar ovens. A century later, Augustin Bernard Mouchot invented the first solar cooker. In 1875, he presented a study on a pot heated by a cylindrical reflector. This prototype was able to cook a one-kilogram stew in four hours. Since then, many models of solar cookers have been developed and spread worldwide, providing a natural and effective way to cook without using gas or electricity. Numerous studies have been conducted to analyze the thermal performance of solar cookers. Among these, (JUBRAN & ALSAAD,1991) It presents a mathematical analysis and an experimental study of the performance of a solar cooker with single and double glazing, with and without a reflector. The results show that the cooker's performance improves as the diameter of the cooking vessel increases. It was also observed that variations in the depth of the vessel have no effect on the cooker's performance. Additionally, a larger solar cooker offers better daily performance compared to smaller ones. Insulating materials play a vital role in enhancing the thermal performance of solar cookers. They minimize heat losses and help maintain a uniform temperature distribution inside the device. (Nahar et al.1993) presented a numerical study on a box-type solar cooker using transparent insulation materials with thicknesses of 40 mm and 100 mm. The results showed that the absorber plate temperature reached 117 °C without TIM and 158 °C with TIM, with corresponding heat loss coefficients of 10.3 W/m^2K and 7.13 W/m^2K, respectively. To further improve the absorption of solar radiation by the absorber plate and achieve higher temperatures, mirrors were added in different configurations to direct and concentrate solar radiation onto the absorber surface. (Emad H,2003, P.2651-2663) A new model of a solar cooker is presented, characterized by the use of a set of diffuse flat reflectors to direct radiation toward the lower surface of the absorber plate. With this method, the absorber plate is exposed to solar radiation on both its upper and lower surfaces. The results, obtained under the same operating conditions, show that the absorbers of the box-type cooker and the double-exposure cooker reach temperatures of 140°C and 165°C, respectively. The air temperatures inside the two cookers are 132°C and 155°C, respectively. Additionally, the double-exposure cooker reduces cooking time by approximately 30 to 60 minutes. (EL -Sebaii &Ibrahim.2005) A solar cooker was designed and tested in two configurations: the first equipped with a single cooking vessel and the second with four vessels. The results show that the F_1 values indicate that the cooker can be used twice a day for successive cooking sessions, while the second merit factor, F_2, demonstrated an almost linear increase with the mass of the cooking fluid. There are different types of solar cookers, with the box-type cooker being one of the most common, varying in design and operation. Another well-known type is the parabolic solar cooker, which concentrates solar radiation onto a single focal point where the cooking vessel is placed. Additionally, there is the panel-type solar cooker. A major drawback of this technology is its dependence on solar energy; that is, during sunset or in the absence of sufficient solar radiation, the solar cooker cannot function effectively. For this reason, many studies focus on integrating thermal storage materials to extend the cooking time of solar ovens. Several studies in the literature have explored the coupling of solar cookers with different thermal storage materials to enhance their efficiency and usability. The addition of reflectors plays an important role in improving the concentration of flux at the absorber of solar cookers. Their principle is based on redirecting the incident rays toward the mirrors; in order to concentrate them on the glazed surface, (Weldu, et al.,2019) They conducted an experimental study on a box-type solar cooker equipped with different reflectors in Bahir Dar., Ethiopia. The maximum plate temperature of 148.7 °C was achieved with a tracking reflector at the optimal angle. Performance, evaluated using thermal efficiency, cooking power, and figures of merit F1 and F2, was significantly improved by using a tracking reflector and an aluminum cooking vessel, showing that optimal solar reflection and good thermal diffusivity of the vessel are key to better cooker performance. The use of high-quality insulating materials and the addition of reflectors help improve the concentration of solar rays, thereby increasing the temperatures inside solar cookers. However, the main drawback remains the intermittency of the solar energy source: its absence at night or during cloudy periods limits the continuous use of these devices. This highlights the need to consider thermal storage materials capable of storing solar energy for later use, (Saxena et al.,2020) A solar cooker coupled with thermal storage materials was designed and tested. Experiments were conducted using three different thermal storage materials. The results show that the use of thermal storage materials significantly enhances the cooker's performance compared to a conventional box-type solar cooker (CSB). Specifically, the cooker achieved a higher efficiency of 53.81%, a lower overall heat loss coefficient of 5.11 W/m^2°C, a higher cooking power of 68.81 W, and an improved heat transfer coefficient of 56.78 W/m^2°C. (ANILKUMAR, B. C. et al, 2021) This study aims to improve box-type solar cookers with thermal energy storage using phase change materials (PCMs) to provide heat after sunset. Various PCMs were evaluated, and erythritol was identified as the best option. The study also determined the optimal PCM mass and storage unit size, showing that erythritol requires the least amount while ensuring efficient operation. It is therefore recommended as the optimal PCM for solar cookers with thermal storage. (KULDHAR & SHRIVASTAVA,2023) A box-type solar cooker with a thermal storage material was designed and tested. The results show that this cooker, coupled with a storage unit, can be used for cooking in the late evening. It allows the phase change material (PCM) to reach temperatures of up to 175°C and offers the possibility of cooking even in the shade. A box-type solar cooker was designed, but this time, the cooker was integrated with a trapezoidal duct. A 200 W halogen lamp was placed inside the duct to enhance heat transfer. Additionally, 450 small hollow copper balls were used to improve the thermal performance of the solar box cooker (SBC). The goal of this modification in the solar cooker configuration was to improve the heat transfer rate and reduce the cooking time while minimizing thermal energy consumption. Experimental results show that the thermal efficiency of the SBC was measured at 45.11%, the cooking power was estimated at 60.20 W, and the overall heat loss coefficient was evaluated at approximately 6.01 W/m^2°C. (ZHOU, Chang, et al, 2023) This study presents an innovative solar cooking system that uses a rock-bed thermal storage to retain heat. Heat is transferred from the solar collectors to the rocks using thermal oils and is then used for cooking food. The results show that the rock-bed storage is the most efficient, as it improves annual performance and reduces the initial cost. Economic analysis indicates that the cost of cooking energy is low, and the system significantly reduces carbon emissions compared to electric or gas cookers. (KULDHAR,Santosh & SHRIVASTAVA,Ritu, 2023) This study presents a box-type solar cooker with inclined sidewalls, designed to enhance thermal performance by reflecting sunlight onto the absorber. The sidewalls are covered with reflective aluminum sheets. The results show that the optimized cooker achieves higher temperatures than a conventional cooker, with a maximum plate temperature of 76 °C versus 65 °C, a maximum pot temperature of 86 °C versus 60 °C, and improved figures of merit F1 and F2.

This paper aims to evaluate the performance of a conventional box-type solar cooker through a comparative thermal study using the thermal simulation software Ansys Fluent, version 2020 R1. Specifically, it analyzes the differences between single-glazing and double-glazing configurations for various types of solar cookers, such as rectangular, trapezoidal, and triangular models, with and without load. Additionally, a modification was made to the conventional rectangular cooker by tilting the internal mirrors at different angles to highlight the impact of this inclination on the solar cooker's thermal efficiency. These simulations were conducted with both single and double glazing, with and without load. The primary objective of this study is to demonstrate the effect of double glazing on the thermal performance of solar cookers.

2 METHODOLOGY

Solar cookers are devices that collect solar radiation at the absorber level and convert it into heat. Once the sun's rays penetrate the glass surface of the cooker, part of this radiation passes through the glass and reflects off the cooker's internal mirrors, which are made of aluminum. These rays are then refracted onto the cooker's absorber, which is painted black to maximize the absorption of solar radiation. To reduce convection losses in the solar cooker, an insulating layer, such as glass wool or wood wool, is integrated.

In this study, different solar cooker geometries are simulated, each with a distinct shape. First, the simulation is performed for the rectangular configuration, where the receiving surface is horizontal with a height of Hp=35 cm. Next, the receiving surface is tilted by an angle β\betaβ to create a trapezoidal configuration with a height of Hp=10 cm. Finally, the triangular configuration is considered, characterized by the largest receiving surface among the three, with a height of Hp=0 cm.

The primary goal of adjusting the front wall height of the solar cooker is to minimize shading on a large portion of the absorber surface. Reducing the front wall height increases the radiation concentration on the absorber, thereby raising the temperature inside the cooker. The choice of the overall cooker height H=35 cm, is based on previous studies reported in the literature. Several researchers have examined solar cookers with different front wall heights. For example (NAHAR, N. M. et al, 1990) performed both experimental and numerical studies on a cooker with a 20 cm height, while B. A. Jubran et al. (JUBRAN, B. A. et ALSAAD, M. A.,1991) designed and tested a solar cooker with a reflector at H′=25 cm. Similarly (NAHAR, N. M., MARSHALL, R. H & BRINKWORTH, B. J., 1994) investigated a model with a height of 27 cm, (AMER, Emad H.,2003) studied a 30 cm version, and (UHUEGBU, C. C.,2011) developed a cooker with a 40 cm front wall. In this study, the rectangular solar cooker was selected with a height of H′=35 cm. To analyze the effect of the inclination of the cooker's internal mirrors on thermal performance improvement, a rectangular geometry is modified by tilting the internal mirrors at different angles α while maintaining the same receiving surface dimensions as the horizontal rectangular solar cooker. This allows for a comparative study of the different cooker geometries in terms of the temperature reached at the absorber surface.

Before designing the solar cookers using SolidWorks, precise dimensions were chosen for each configuration. All configurations consist of an absorbing plate painted black to maximize solar radiation absorption, as black color is known for its high absorption capability. The absorber plate is made of aluminum with a thickness of 1.5 mm, and the mirrors used have a thickness of 1 mm.To create the greenhouse effect, two glass layers with a thickness of 3 mm are used. The chosen dimensions of the cookers are such that the width is 400 mm, the length is 535 mm, and the height varies for each configuration: H = 35 cm for the rectangular case, H = 10 cm for the trapezoidal case, and H = 0 cm for the triangular case.

For the modified rectangular case with inclined internal mirrors, different inclination angles are considered: $\alpha = 45°$, $\alpha = 75°$, and $\alpha = 89°$, while keeping the other dimensions identical to the configurations studied. The design of the different solar cooker geometries was carried out using SolidWorks and then imported into Ansys Fluent to perform thermal simulations. Figure (1) illustrates all the configurations chosen for this study along with their dimensions:

Figure 1. Trapezoidal solar cooker designed by SolidWorks

Hg=35cm

Lext=53,5cm

Wext =40 cm

eg=0,3cm

e_{iso}=2,2cm

e_W=4cm

Hp =10cm

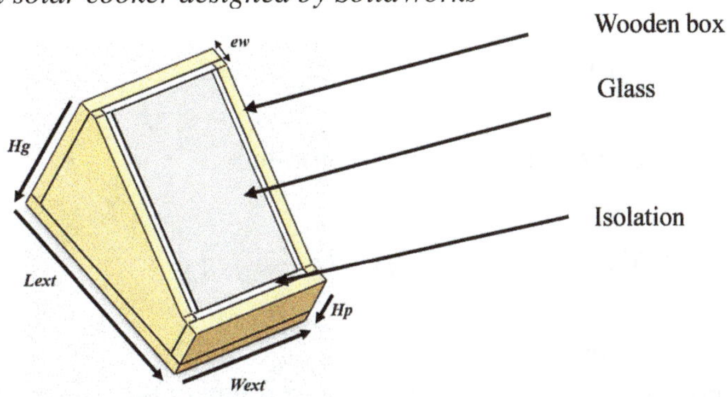

The figure (1) shows the trapezoidal solar cooker designed by SolidWorks with its dimensions. The difference between the various configurations studied in this article lies in their height. The rectangular configuration has a height of Hp= 35 cm, the second geometry has a height of Hp= 10 cm, and the third configuration is the triangular one with a height of Hp= 0 cm. The purpose of reducing the height of the solar cooker's front wall is to maximize the absorption of solar radiation. By decreasing the height, part of the shadow cast on the absorber surface is reduced, thereby enhancing the solar cooker's performance.

3 CFD METHODE

After designing and importing the different solar cooker geometries into the Ansys Fluent software, an analysis of their performances was conducted using the CFD method. This approach involves solving fluid dynamics phenomena by solving the Navier-Stokes differential equations. The aim of this simulation is, initially, to compare a box-type solar cooker with single glazing and another with double glazing, focusing on the absorber plate temperature. Then, the study examines the impact of the load on the absorber's temperature and its evolution for both configurations (single and double glazing). The CFD method relies on solving fluid flow problems in a defined domain by applying the Navier-Stokes equations. This approach involves dividing the fluid domain into thousands of small elements called meshes. The equations used to solve these problems include the mass conservation equation, the momentum equation, and the energy equation, which are presented as follows (1)-(2)-(3).

Equation of continuity

The equation of mass conservation is written in the form

$$\frac{\partial \rho}{\partial t} + \vec{\nabla}(\rho * \vec{v}) = 0 \tag{1}$$

Equation of conservation of momentum

$$\frac{\partial(\rho * \vec{v})}{\partial t} + \vec{\nabla} . (\rho * \vec{v} * \vec{v}) = \vec{\nabla} * \vec{p} + (\rho - \rho_0) . \vec{g} \tag{2}$$

Energy equation

For an incompressible fluid with constant conductivity, the energy equation is written in the form:

$$\rho C_p \frac{dT}{dt} = \lambda \nabla^2 T + \mu \phi \tag{3}$$

The CFD method involves generating a mesh for the corresponding geometry based on its characteristics. The role of the mesh in Ansys Fluent is crucial, as it influences both the stability and the computation time of the simulations. The mesh divides the fluid elements to be analyzed into thousands of finite elements, called cells, in order to solve the Navier-Stokes equations. The accuracy of the solution and the efficiency of the simulation largely depend on the quality and refinement of the mesh, as finer meshes allow for more accurate results, though at the cost of longer computation times.

In our case, a mesh test was conducted for the three types of solar cookers selected for this study in order to choose an appropriate number of elements that would ensure the stability of the simulations while reducing computation time. The number of mesh elements was tested for different values: 23,800, 30,600, 36,300, 46,200, 61,600, 113,660, and 165,840, to ensure the stability and convergence of the calculations. These tests were performed for the different geometries of the selected solar cookers: the rectangular case with a horizontal receiver surface of height H = 35 cm, the second configuration being the trapezoidal case with an inclined receiver surface, and the third geometry, the triangular case, characterized by the absence of the front wall to improve solar radiation concentration. Figure (2) presents the results of the variation in absorber temperature as a function of the number of elements in Ansys Fluent for the three solar cookers.

Figure 2. The evolution of the absorber temperature as a function of the number of elements for the three solar cookers: (a) Triangular, (b) Trapezoidal, (c) Rectangular

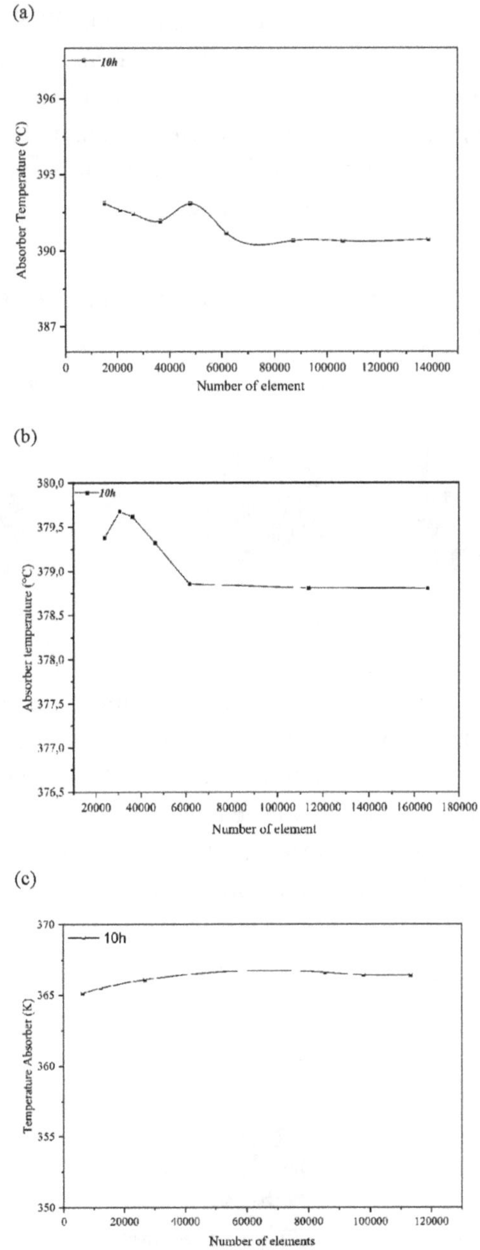

The three mesh tests conducted for the different geometries of the solar cookers allowed us to analyze the impact of the number of elements on the accuracy of the results and the stability of the simulations. As illustrated by the results presented in Figure (2), it was observed that beyond 90,000 to 100,000 elements, the absorber temperature stabilizes and shows no significant variations. This indicates that adding more elements does not bring any notable improvement in terms of accuracy, but it does increase the computation time. Therefore, to optimize the simulations, a balance was found by selecting a sufficient number of elements to ensure the accuracy of the results while reducing the simulation time.

Figure 3. Meshing for solar box cooker (a) Rectangular, (b) Triangular, (c) Trapezoidal without load

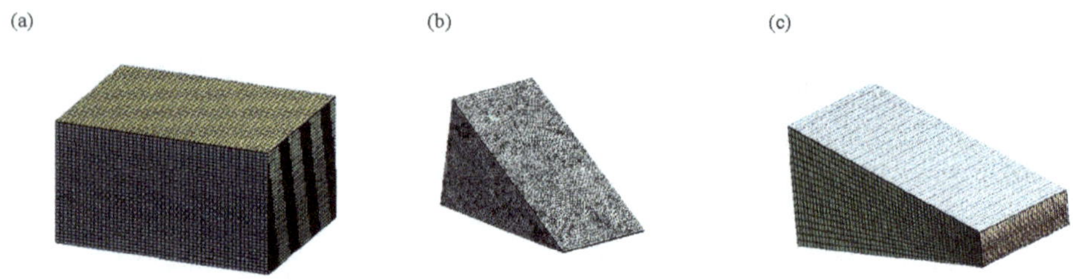

In the first simulation without load, a structured mesh was used, as shown in the figure. This choice ensures greater accuracy and promotes faster convergence of calculations. However, in the simulation with load, which includes a cylinder filled with water and an air zone, the geometric complexity of the domain increases. In this case, a tetrahedral mesh is preferred, as it better adapts to complex shapes while ensuring numerical stability and good solution convergence.

Figure 4. Meshing for the three types of solar cookers with load (a) triangular. (b) Trapezoidal, (c) Rectangular

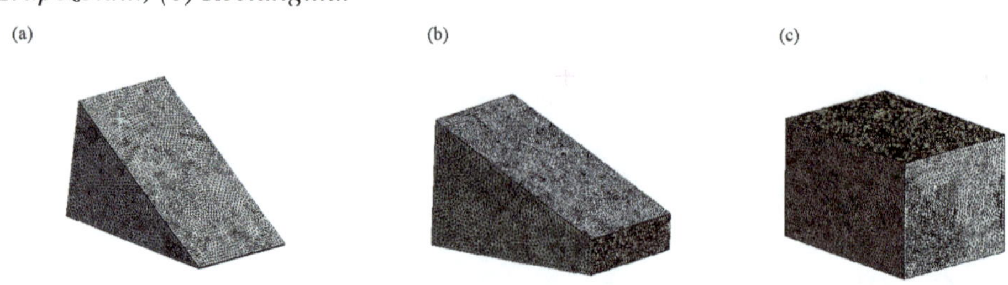

After the meshing step and the selection of the appropriate number of elements, it is essential to assign the corresponding material to each component of the solar cooker and define the boundary conditions suited to our study. These parameters directly influence the accuracy and reliability of the results obtained. Regarding turbulence modeling, Ansys Fluent offers several models, primarily distinguished by the number of equations used to solve the complex physical phenomena of turbulent flows. The Spalart-Allmaras (SA) model solves a single transport equation for an auxiliary variable related to turbulent kinematic viscosity, making it computationally efficient and suitable for free-flow and low Reynolds number applications. The Reynolds Stress Models (RSM) are more advanced and directly solve transport equations for Reynolds stresses, making them better suited for highly anisotropic flows, though they are more computationally expensive. The two-equation models, including the k-ε and k-ω models, solve transport equations for turbulent kinetic energy (k) and its dissipation rate (ε) or the specific dissipation frequency (ω). Among the k-ε variants, the standard k-ε model introduced by Launder and Spalding (Jones & Launder,1972) is one of the most commonly used due to its robustness, low computational cost, and suitability for fully developed turbulent flows. This model assumes negligible molecular viscosity and is widely applied for its stability and efficiency in numerical simulations (ATTOU,2019) – (KHALDI,2018).

In this study, the standard k-ε model was chosen for its ability to provide a good balance between accuracy and computational cost, making it particularly suitable for simulating both natural and forced convection in solar cookers. The transport equations in This model are as follows:

$$\frac{\partial}{\partial t}(\rho K) + \frac{\partial}{\partial x_i}(\rho K u_i) = P_K - \rho\varepsilon + \frac{\partial}{\partial x_j}\left(\Gamma_k \frac{\partial k}{\partial x_j}\right) \tag{4}$$

$$\frac{\partial}{\partial t}(\rho\varepsilon) + \frac{\partial}{\partial x_i}(\rho\varepsilon u_i) = C_{\varepsilon 1}\frac{\varepsilon}{K}P_K - C_{\varepsilon 2}\rho\frac{\varepsilon^2}{K} + \frac{\partial}{\partial x_j}\left(\Gamma_K \frac{\partial\varepsilon}{\partial x_j}\right) \tag{5}$$

$$\Gamma_K = \mu + \frac{\mu_t}{\sigma_K} \text{ and } \Gamma_\varepsilon = \mu + \frac{\mu_t}{\sigma_\varepsilon}$$

P_K: The production of turbulent kinetic energy due to the gradient is defined by the following equation

$$P_K = \mu_t\left(\frac{\partial U_i}{\partial X_j} + \frac{\partial U_j}{\partial X_i}\right)\frac{\partial U_i}{\partial X_j} + \frac{2}{3}*\rho*K\delta_{ij}\frac{\partial U_i}{\partial X_j} \tag{6}$$

For the standard k-ε model, the constants are defined by the following values, which are typically used in simulations:

$C_{\varepsilon 1}$ = 1,44 Constant for the transport equation of k
$C_{\varepsilon 2}$ = 1,92 Constant for the transport equation of ε
C_{μ} = 0,09 Constant associated with turbulent viscosity
σ_K = 1,0 Prandtl number for k (represents the diffusion scale for k)
σ_{ε} = 1,3 Prandtl number for ε (represents the diffusion scale for ε)

These constants are used to solve the transport equations for k (turbulent kinetic energy) and ε (the dissipation rate of turbulent kinetic energy), and they play a crucial role in the stability and accuracy of simulations in turbulent flows.

3.1 The standard k-ω Model

The standard k-ω model also solves two equations, requiring a very fine mesh near the wall. The first equation concerns the turbulent kinetic energy k, and the second equation corresponds to the specific dissipation rate ω\omegaω. These two equations are defined as follows:

$$\frac{\partial(\rho k)}{\partial t} + \frac{\partial(\rho\ U_j K)}{\partial X_j} = P_K - \beta'\rho\ KW + \frac{\partial}{\partial X_j}\left(\Gamma_K \frac{\partial K}{\partial X_j}\right) \tag{7}$$

$$\frac{\partial(\rho W)}{\partial t} + \frac{\partial(\rho\ U_j\ W)}{\partial X_j} = a\ \frac{W}{K}\ P_K - \beta\ \rho\ W^2 + \frac{\partial}{\partial X_j}\left(\Gamma_W \frac{\partial W}{\partial X_j}\right) \tag{8}$$

$$\Gamma_K = \mu_t + \frac{\mu_t}{\sigma_t}\ \Gamma_W = \mu + \frac{\mu_t}{\sigma_W}$$

The constants for the standard k-ω model are defined as follows:

β' =0.09 α=5/9 β =3/40 W_K=2.0, σ_W=2.0

These constants help define the behavior of the turbulence model, particularly for the calculation of turbulent dissipation and energy production in regions with high velocity gradients, such as near walls or in concentrated flows.

3.2 Boundary Condition

To ensure accurate results in CFD simulations of solar cookers, it is essential to rigorously define the boundary conditions, which represent the thermal and physical exchanges between the different components of the system. The glazing plays a crucial role by allowing solar radiation to enter the cooker while minimizing heat losses, thereby enabling high temperatures to be reached inside the cooking chamber. When exposed to solar radiation, part of it is transmitted and absorbed by the absorber plate, while the rest is either reflected or absorbed by the glazing itself. This absorption leads to an increase in the interior air temperature, further enhancing the heating of the absorber.

However, heat losses occur due to natural convection between the glazing and the cooker's air, as well as with the surrounding ambient air. Additionally, a portion of the thermal radiation is re-emitted outward, resulting in radiative losses. Implementing double glazing is an effective solution to minimize these losses, as it creates an insulating barrier that reduces thermal exchanges with the environment and enhances the thermal efficiency of the solar cooker.

The absorber plate, typically made of a material with high absorptivity, plays a key role in converting solar energy into heat. Once the incident radiation is captured, the generated heat is transferred to the interior air primarily through convection, helping maintain high temperatures inside the cooker.

Moreover, the external walls of the cooker, being in direct contact with ambient air, experience heat losses through convection and radiation. The extent of these losses depends on various factors, such as wind speed, ambient temperature, and the insulating properties of the materials used. Effective thermal insulation of the walls helps reduce these losses and optimizes heat retention, thereby improving the overall performance of the solar cooker.

Therefore, a precise definition of boundary conditions in Ansys Fluent is crucial for accurately modeling the thermal behavior of the cooker and obtaining reliable simulation results.

The equations defining these phenomena are presented below:

In Ansys Fluent, the radiation and convection mechanisms are modeled using the "Mixed" option and are combined through the following equation:

$$-\lambda_v \frac{dT}{dn} = h_0\left(T_v - T_a\right) + \varepsilon_v \sigma\left(T_v^4 - T_{SKY}^4\right) \tag{9}$$

With T_{SKY} is the sky temperature calculated as follows (SWinbank,1963):

$$T_{SKY} = 0.0552 * T_a^{1.5} \tag{10}$$

Regarding the wooden walls and the absorber surface, they are subjected to a convection heat loss, and the heat transfer coefficient is given by the following relation (AKOY & AHMED,2015):

$$h_0 = 2,8 + 3*V \tag{11}$$

Based on the mesh treatment, activation of the energy equations, and selection of the turbulence model, to begin the simulation in Ansys Fluent, it is necessary to define the materials for each part of our geometry. The different materials used with their thermal properties are presented in the following table:

Table 1. Thermophysical Properties.

Materials	Density ($Kg.m^{-3}$)	Specific Heat ($J. Kg^{-1}. K^{-1}$)	Thermal conductivity ($W. m^{-1}. K^{-1}$)
Aluminium	2719	871	202,4
Wood	600	1600	0,12
Glass	2530	720	1,2
Air	1,225	1005	0,0262
Water	997	4182	0,6

In ANSYS Fluent, discretization is a crucial step for solving the conservation equations (mass, momentum, energy). It involves transforming these differential equations into algebraic equations that can be solved on a discrete mesh. The choice of discretization schemes in ANSYS Fluent impacts the results, stability, and convergence of the solutions. ANSYS Fluent offers several discretization schemes. The following table presents the discretization scheme used in our simulation:

Table 2. Discretization schemes used in ANSYS-FLUENT simulation

Spatial discretization	Scheme
Gradient	Least squares cell-based
Pressure	PRESTO
Momentum	Second-Order Upwind
Turbulent Kinetic Energy	Second-Order Upwind
Turbulent Dissipation rate	Second-Order Upwind
Energy	Second-Order Upwind

To ensure good calculation accuracy, it is essential to properly define the under-relaxation parameters, which play a fundamental role in the stability and convergence speed of the simulations. These parameters regulate the fraction of change in variables (pressure, velocity, temperature, turbulence, etc.) applied at each iteration. An inappropriate choice of these values can lead to calculation divergence. In our case, for the simulations performed, the under-relaxation parameters are defined as follows: 0.3 for density, 0.1 for body forces, 0.4 for turbulent kinetic energy, 0.4 for turbulent dissipation rate, 0.5 for turbulent viscosity, and 0.7 for energy

4 CFD SIMULATION FOR DIFFERENT TYPES OF SOLAR COOKERS

The glazing surface of solar cookers plays a crucial role in the accumulation of solar radiation inside the solar cooker. When a ray strikes the receiver surface of the cooker, part of it is reflected off the glass surface due to the change in the refractive index between air and glass. Another portion is transmitted, meaning it passes through the glass into the interior of the cooker, where it is trapped and converted into heat to increase the temperature.

The purpose of this simulation is to conduct a comparative thermal study of the absorber temperature for two cases: one with single glazing and one with double glazing, in order to determine the effect of double glazing on the thermal performance of solar cookers with different geometries. The simulation is carried out for a typical summer day in July, without any load, using a single 3mm thick glass for all three types of selected solar cookers. Then, a 1-liter water load is added to each of the three cookers. After that, another layer of glazing is added to these three types of solar cookers to rerun the simulation with and without the load to determine the necessary parameters for analyzing the thermal performance of the cookers.

Additionally, another simulation will be performed for the modified rectangular configuration, where the internal mirrors are tilted at various angles, with and without the load, and with both single and double glazing. The goal of these different simulations is to make a comparison of the thermal performance of the solar cookers by calculating several key parameters.

4.1 First Figure of Merit

The first figure of merit is calculated for solar cookers exposed to the sun, without load and without a reflector, to determine their classification. It is defined as the ratio of optical efficiency to the heat loss coefficient (IBRAHIM & EL-REIDY,2007):

$$F_1 = \frac{T_P - T_a}{H} \tag{12}$$

4.2 Loss Heat

The efficiency of solar cookers mainly depends on their ability to minimize thermal losses to achieve maximum temperatures inside the device. This coefficient can be calculated using the correlation provided in equation (13), Various parameters can be used to reduce this coefficient and enhance the performance of solar cookers, such as the use of reflectors, the tilting of the cooker's mirrors, and the increase in insulation thickness. These adjustments help minimize thermal losses and optimize solar energy absorption, thereby achieving higher temperatures inside the cooker (CHANNIWALA & DOSHI, 1989) – (ADEWUMI et al.,2019)

$$Q_t = \frac{\left(T_p - T_a\right)}{\frac{N}{\left(\left(\frac{C}{T_p}\right)*\left(\frac{\left(T_{p-T_a}\right)}{(N-f)}\right)^{0.33}\right)} + \frac{1}{h_0}} - \frac{\sigma*\left(T_p^4 - T_a^4\right)*A_p}{\frac{1}{\left(\varepsilon_p + 0,05*N*\left(1+\varepsilon_p\right)\right)} + \left(\frac{2*N+f-1)}{\varepsilon_g}\right) - N} \tag{13}$$

Where is the top loss coefficient, Ut can be calculated by:

$$U_t = \frac{Q_t}{A_p*\left(T_p - T_a\right)} \tag{14}$$

U_t: loss heat in (Wm^{-2} °C)
N: the number of glass layers used
T_a The ambient temperature in °C, T_P: The absorber temperature in °C

f=(1-0.04*H_{wind}+0.0005*H_{wind}^2) (1+0.091N)(15)

$H_{wind:}$ the heat transfer coefficient (W/m^2 C) is defined by

H_{wind}=2.8+3*V (16)

V is the wind speed in m/s.

C=365.9*(1-0.00883β+0.0001298$β^2$) (17)

ε_p: emittance of plate, ε_g: emittance of glass, h_0: Heat transfer coefficient, A_P: is the plate area (m^2), β: the angle of inclination of the glass; in our case, since we have a box-type solar cooker, we take β=0 for rectangular configuration, β=25,04 for trapezoidal and β=33,192 for triangular configuration.

4.3 Second Factor of Merit

This factor is determined when the cooker is under load and evaluates the efficiency of the solar cooker in converting incident solar energy into useful heat. It can be expressed by the Following equation:

$$F_2 = \frac{F_1*(M_w*C_w)}{A*t} * \ln\left(\frac{1-\frac{1}{F_1}*\left(\frac{T_{w1}-T_a}{H_m}\right)}{1-\frac{1}{F_1}*\left(\frac{T_{w2}-T_a}{H_m}\right)}\right) \quad (18)$$

Tw1 is the initial water temperature (°C), T_{W2} is the final water temperature (°C), A is the aperture area (m^2), where t is the time interval (s), during which water temperature rises from T_{w1} to T_{W2}. I and Ta are the average horizontal solar radiation (W/m^2) and ambient temperature (8C) over the time interval t, Ta (°C) the average ambient temperature from the time period t1 to t2, G (W/m2) the average solar radiation from the time period t_1 to t_2 and M_W mass of water C_W is the specific heat capacity of water (J/kg °C),

After determining F_1 and F_2 the boiling time (t), required to evaluate sensible heat, is determined using the following equation:

$$t = -\frac{F_1*M_W*C_W}{F_2*A} \ln\left(1 - \frac{T_{W2}-T_{am}}{F_1*g}\right) \quad (19)$$

The boiling time of the load is determined based on the two figures of merit F_1 and F_2, as well as the evaluation of the final water temperature T_{W2} and the ambient temperature Tam of the surrounding environment.

4.4 Overall Thermal Efficiency of the Solar Box Cooker

It is a performance indicator coefficient for the solar cooker, representing the ratio between the useful thermal energy transferred to the cooker's contents and the incident solar energy captured by the cooker. It is defined by the Following equation (Weldu et al., 2019):

$$\eta_0 = \frac{M_W*C_W*\Delta_T}{H*A*\Delta_t} \quad (20)$$

M_W is mass of water (kg), C_W is specific heat capacity of water (J/kg/K), ΔT is temperature difference between maximum temperature of water and the ambient temperature (°C), A is area of aperture of the cooker (m^2), Δt is time required to achieve the maximum temperature of cooking, H is Average solar intensity during the interval Δt (W/m^2).

The cooking power (KUMAR et al.,2008) can be defined as the rate of useful energy available during the heating period can be calculated by relation:

$$P = Mw*Cw*\frac{(T_{W2} - T_{W1})}{600} \tag{21}$$

This equation has been divided by 600 to account for the number of seconds in each 10-minute time interval.

5 RESULTS AND DISCUSSIONS

The objective of the simulation performed using the CFD method for the selected types of solar cookers is to analyze their performance with and without load, as well as with single and double glazing. The aim is to evaluate the effect of double glazing on the thermal performance of the solar cookers and its impact on reducing cooking time. The various simulation results are presented for both cases: with and without glazing, to determine the first figure of merit, and with a load for both single- and double-glazing configurations across the different selected solar cookers. This allows for the assessment of the second figure of merit and the analysis of the overall thermal efficiency of the solar cookers.

5.1 Case Without Load and With Simple Glass and Double Glass

All simulations were conducted for a typical day in July, characterized by maximum solar irradiation. Initially, the simulations were performed on rectangular, trapezoidal, and triangular solar cookers without load and with a single glazing. Then, they were repeated without load but with double glazing to determine the air temperature inside the cooker and the stagnation temperature of the absorber. The obtained results are presented in Figure (5):

Figure 5. Evolution of the absorber temperature for the rectangular configuration (H = 35 cm), trapezoidal configuration (H = 10 cm), and triangular configuration (H = 0 cm) with single and double glass

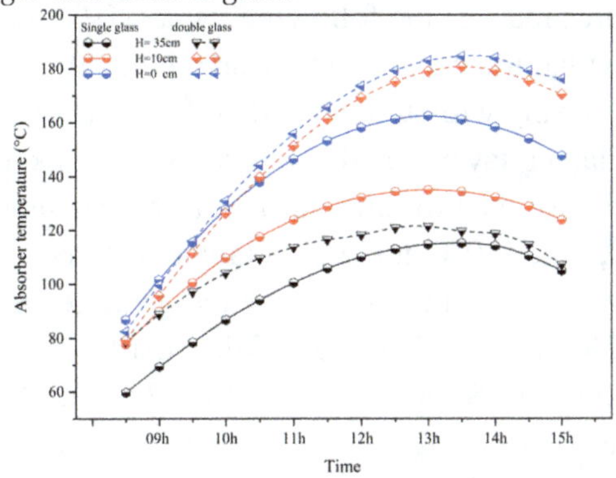

Figure (5), shows that the absorber temperature and the interior air temperature of the solar cooker decrease as the height of the cooker increases. It is observed that the triangular configuration exhibits the highest absorber and interior air temperatures compared to the trapezoidal and rectangular cases. The stagnation temperatures of the absorber, for the no-load case with single glazing, reach 112°C, 135°C, and 160°C at the absorber surface, and 80°C, 90°C, and 110°C for the interior air temperature of the cooker, respectively, for the rectangular, trapezoidal, and triangular configurations,

Figure 6. Evolution of the indoor air temperature for the rectangular configuration (H = 35 cm), trapezoidal configuration (H = 10 cm), and triangular configuration (H = 0 cm) with single and double glass

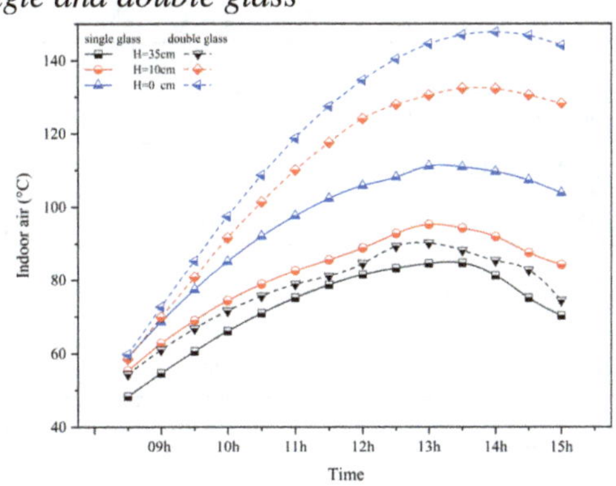

To analyze the performance of the solar cookers, it is essential to calculate their first figure of merit in order to classify them based on their operating conditions. By applying equation (), the first figure of merit corresponding to the absorber stagnation temperatures reaches the following values: 0.08, 0.099, and 0.1246 °C m^2 /W for the rectangular, trapezoidal, and triangular configurations, respectively, in the case of single glazing without load and without a reflector.

When a second glazing layer is added to each solar cooker, an increase in absorber temperature is observed for all configurations. The absorber temperatures reach 126°C, 170°C, and 180°C for the rectangular, trapezoidal, and triangular cases, respectively. Similarly, the interior air temperature of the cooker reaches 96°C, 120°C, and 138°C. The calculation of the first figure of merit under these conditions gives values of 0.090, 0.1349, and 0.1452 °C m^2/W for the rectangular, trapezoidal, and triangular configurations, respectively, in the case with double glazing, without load, and without a reflector.

According to the Indian Bureau of Standards (BIS), for a solar cooker to be effective according to these standards, both the first and second merit factors (F1 and F2) must meet the following criteria: F1 > 0.11 cm^2/W and F2 > 0.4 (Al-NEHARI et al.,2021) – (HARMIM et al., 2012) Based on the values of F1 obtained, we can classify the solar cookers as follows: the rectangular solar cooker falls into class B because its first merit factor is less than 0.11, while the trapezoidal and triangular configurations can be classified as type A because their first merit factors are greater than 0.11 (Al-NEHARI et al.,2021) - (HARMIM et al., 2012). The following table presents some first merit factor values found in different studies:

Table 3. Comparison of the experimental findings with the new studies

Reference	Maximum stagnation temperature (°C)	First figure of merit F_1 (°C m^2 /W)
(KUMAR et al.,2008)	140	0,1681
(SHRESTHA & BYANJANKAR, 2007)	120	0.19
(EKECHUKWU & UGWUOKE,2003)	*110*	*0,1*
(ADEME & HAMEER,2018)	*161,7*	*0,123*
(WASSIE et al.,2022)	91	0.08
(ISHAQ et al.,2023)	*119*	*0.099*
(EL-SEBAII & IBRAHIM,2005)	*101*	*0,15*
(KUMAR et al.,2008)	138,46	0,117
(MILIKIAS et al.,2021)		*0.115*
(SAXENA & KARAKILCIK,2017)	*94*	*0.13*

When a second glazing is added, an air gap forms between the two panes, acting as a thermal insulator. This space limits heat exchanges between the inside and outside of the solar cooker, thereby reducing thermal losses, particularly through convection. This insulation helps improve the energy efficiency of the cooker by maintaining a higher temperature at the absorber, as confirmed by the obtained results. Additionally, the addition of a second glazing enhances the greenhouse effect generated during the solar cooker's operation. By trapping a greater amount of infrared radiation re-emitted by the absorber, it helps minimize thermal losses and maximize heat accumulation inside the device. Thus, this configuration optimizes the thermal efficiency of the cooker and enhances its performance in solar cooking.

Figure 7. Temperature distribution for the three types of solar cookers with double glazing; (a) triangular, (b) Trapezoidal, (c) Rectangular

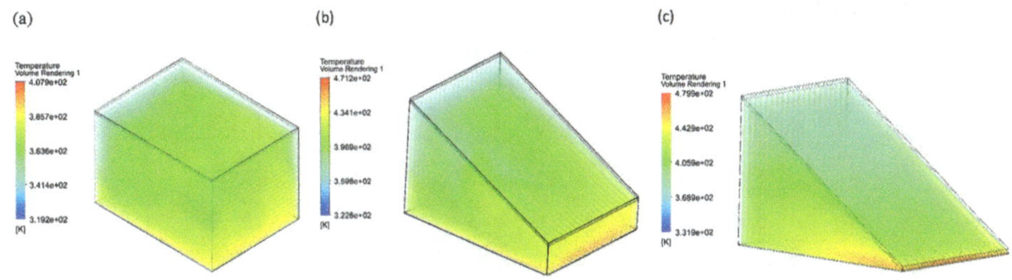

The figures illustrate the temperature distribution in the three types of solar cookers selected for this study. They show an increase in temperature in the triangular cooker compared to the trapezoidal and rectangular configurations. This temperature rise is highlighted by the maximum values reached on the absorber surface. The results indicate that the trapezoidal and triangular configurations offer better thermal performance compared to the rectangular model. Therefore, to enhance the thermal efficiency of the rectangular cooker, it would be beneficial to add reflectors or tilt the internal mirrors.

5.2 Case With Load and With Simple Glass

In this section, a thermal load was applied to the three solar cookers, placed inside an aluminum cylindrical enclosure with a diameter of 18 cm and a height of 10 cm, equipped with a glass lid. This configuration allows for the study of the impact of adding thermal load on the temperature of the absorber. The direct contact between the solar cooker's surface and the base of the cylinder promotes heat exchange between these two surfaces. This heat transfer influences the thermal distribution within the cooker and alters the temperature evolution of the absorber. Thus, this thermal interaction plays a key role in the system's efficiency. It enables the analysis of the effect of adding thermal load on the temperature evolution and the thermal efficiency of the solar cooker. The goal of this study is to determine the impact of adding a second glazing and the role of the air gap created between the two panes on the cooking time of solar cookers to reach the maximum water temperature, as well as on the reduction of heat losses. A solar cooker is considered efficient if the cooking time is reduced and higher temperatures are reached in a shorter period. The following figure shows the temperature evolution of the thermal load in the three selected types of solar cookers as a function of time for the cooker with load and single glazing, simulated using the CFD method throughout the entire day.

Figure 8. Evolution de la température de l'eau pour les deux cas trapezoidal et triangulaire with simple glass

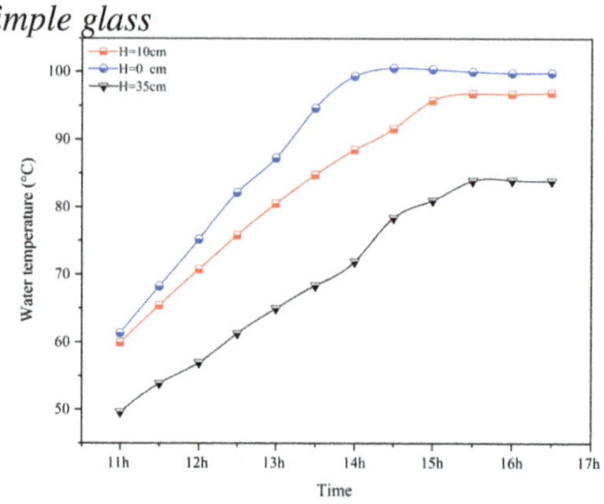

The figure (8) shows the evolution of water temperature in the solar cookers throughout the entire simulation day. It is observed that the temperature of the load in the cookers increases to reach the maximum temperature value at 14:30 for the trapezoidal case, 13:30 for the triangular case, and 15:30 for the rectangular case. Therefore, the effect of the geometry is evident in the reduction of the time it takes for the load temperature to increase in the solar cookers. This can be explained by the fact that the triangular geometry, lacking a front wall, concentrates more solar radiation than the rectangular configuration, due to the absence of shadow on a large portion of the absorber, which helps this geometry concentrate more solar radiation.

Based on the application of the relationship (18) to calculate the second merit factor, which depends on the time taken to reach the boiling temperature of the water and its temperature, the calculation reveals that the second merit factor reaches values of 0.1872, 0.2150, and 0.3609, respectively, for the rectangular, trapezoidal, and triangular cases. The second merit factor depends on the increase in water load temperature from T_{W1} to T_{W2}. If this period is shorter, the second merit factor yields better values; if it is longer, it decreases. This means that a more efficient solar cooker is one that cooks the elements in a shorter time.

5.3 Case With Load and With Double Glass

In this section, the water load was added to the three solar cookers with double glazing to observe their effect on the boiling time of the water load, comparing it with the single glazing case. The simulation was performed for the same selected day for this study. The following figure shows the simulation results for the selected day of this study:

Figure 9. Evolution of water temperature for the trapezoidal and triangular solar cooker with double glazing

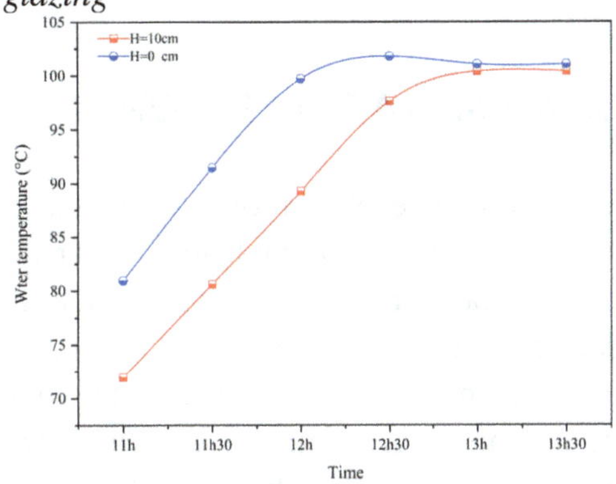

According to Figure (9), which presents the simulation results for the three cases with double glazing and the same water load previously used in the first study with single glazing, it is observed that for all three cases, the boiling time of the water is reduced after adding the second glass. This addition helps to reduce heat losses, thereby improving and increasing the temperature inside the cooker. Based on the calculation of the second figure of merit, the following values are obtained: 0.5211, and 0.675 for the trapezoidal, and triangular cases with double glazing, respectively. These values correspond to the values proposed by the BIS (Al-NEHARI et al.,2021) – (HARMIM et al., 2012),

Figure 10. Temperature contours for solar box cooker with double glazing and with load

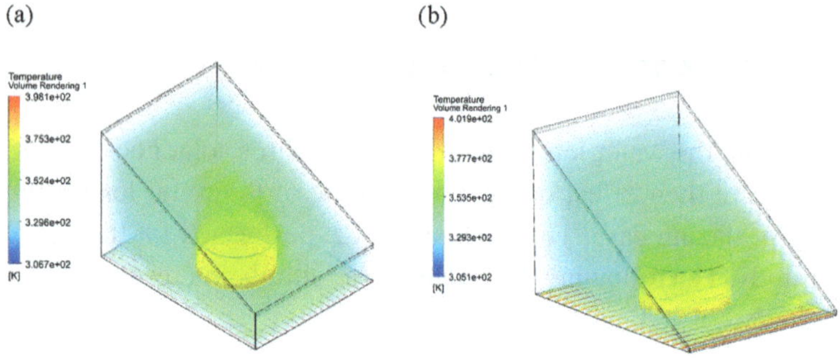

6 THE IMPACT OF ADDING A REFLECTOR

The previous results show the evolution of the absorber and interior air temperature of the solar cookers with and without load. In this part, another simulation will be conducted for the rectangular solar cooker to better demonstrate the effect of adding a main reflector on the thermal performance of the cooker. The reflector helps improve the concentration of solar radiation inside the cooking chamber, which influences the interior air temperature of the cooker. To ensure the reflector works optimally, it is first necessary to calculate the optimal tilt angles of the reflectors, which vary throughout the day. The following equation will be used to calculate the optimal tilt angle of the main reflector for the rectangular solar cooker case (Duffie & Beckman,2013) -(Mirdha & Dhariwal,2008)

$$\varphi' = \frac{2}{3}(\pi + \beta - \theta) \quad (22)$$

β is the tilt angle of the glass receiver surface for the rectangular case, which is equal to zero. The incidence angle θ represents the angle between the sun's rays hitting a surface and the plane normal to that surface, as defined by Duffie and Beckman (Hailemariam M. WASSIE, 2022):

$$cos\theta = cos(\varphi - \beta) \times cos\delta \times cos\omega + sin(\varphi - \beta) \times sin\delta \quad (23)$$

The following figures show the evolution of the absorber and interior air temperature in solar cookers, with and without the addition of a primary reflector, to highlight its effect on temperature variation. This temperature increase is due to the fact that the reflectors redirect an additional amount of incident radiation towards the absorber, thereby increasing the received energy intensity.

Figure 11. Absorber temperature with and without reflectors for solar box cooker H=35cm

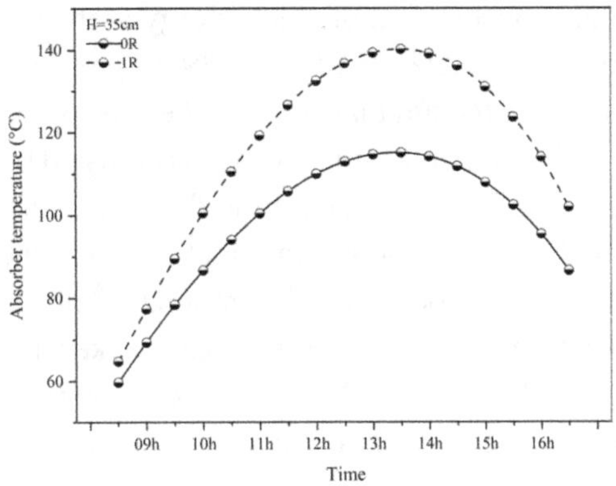

Figure 12. Indoor air temperature with and without reflectors for solar box cooker H=35cm

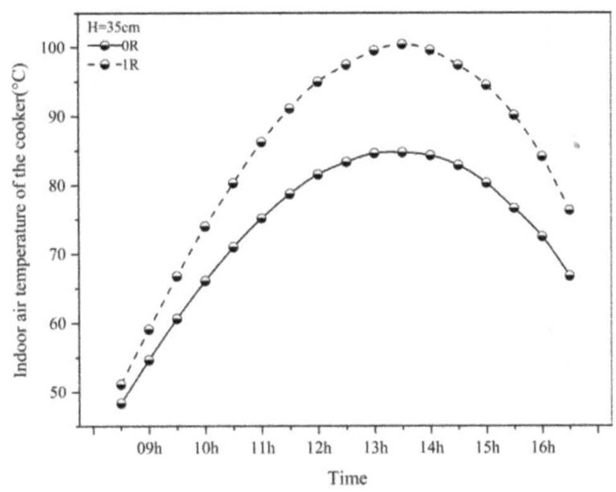

The simulation results for cases with and without the main reflector are presented in Figure (12), clearly illustrating the effect of adding the main reflector on the absorber temperature and the air inside the solar cooker. According to the results obtained, the absorber temperature increases from 112°C to 140°C for the cases without and with the main reflector, respectively. Similarly, the air temperature inside the cooker reaches 85°C without a reflector and 100°C with the addition of the main reflector. Furthermore, the first figure of merit increases from 0.08 °C·m²/W to 0.11 °C·m²/W for the cases without and with the main reflector, respectively. (Hailemariam M. Wassie) tested a box-type solar cooker with and without aluminum and glass reflectors. According to their experimental results, the absorber stagnation temperature reached 161.3°C with a glass reflector, 112°C with an aluminum foil reflector, and 91°C without a reflector. These findings demonstrate that adding reflectors enhances the concentration of solar energy at the absorber level and improves system efficiency by 16.8%, 23.5%, and 39.4%, respectively.

Just like in the case of the rectangular solar cooker, the optimal inclination angle of the main reflector in the triangular solar cooker is adjusted hourly, as the position of the sun continuously changes throughout the day. The calculation of this angle directly depends on both the solar hour angle and the solar inclination angle. The figures (13) illustrate the evolution of the air temperature inside the cooker, as well as the absorber temperature, for the triangular solar cooker in both scenarios: with and without a reflector. This analysis helps assess the impact of the reflector on the thermal performance of the system, particularly in terms of optimizing solar radiation absorption and enhancing heat retention inside the cooker.

Figure 13. Absorber temperature with and without reflectors for solar box cooker H=0cm

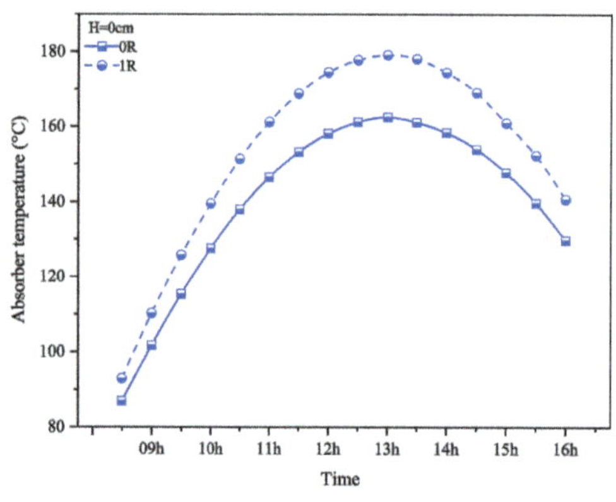

Figure 14. Indoor air temperature with and without reflectors for solar box cooker H=0cm

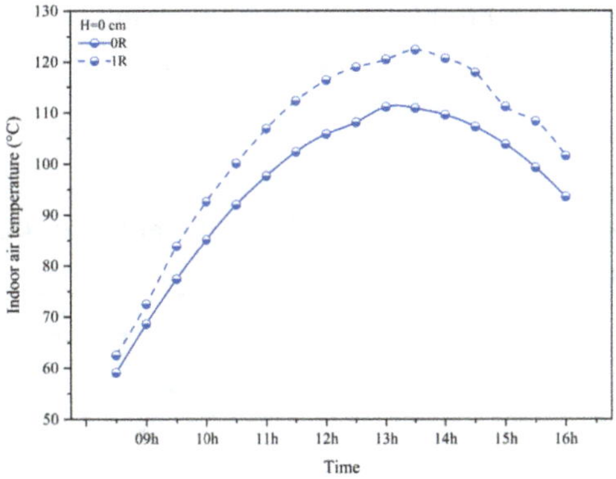

Based on the two figures showing the evolution of the absorber and interior air temperatures of the triangular solar cooker, with and without the main reflector, a significant temperature increase is observed when the reflector is added. The maximum absorber temperature rises from 160°C without a reflector to 180°C with a reflector, while the air temperature inside the cooker increases from 110°C to 122°C, respectively. The addition of the main reflector aims to highlight its effect on the concentration of solar radiation at the absorber level, thereby improving the cooker's thermal efficiency. As a result, the first figure of merit increases after integrating the reflector, reaching 0.145 °C·m²/W compared to 0.1246 °C·m²/W in the case without a reflector.

7 WITH INCLINATION OF INTERNAL MIRRORS

In this section, the internal mirrors of the solar cooker are tilted at different angles relative to the horizontal while maintaining the same receiver surface dimensions as the conventional rectangular solar cooker. The goal of this configuration is to enhance the concentration of solar rays on the absorber.

This improvement is due to the fact that inclined mirrors reflect incident rays more effectively than vertical walls. In a cooker with vertical mirrors, a portion of the solar radiation may be reflected in non-optimal directions, reducing the energy captured by the absorber. Conversely, tilting the mirrors directs more reflected rays toward the absorber, minimizing losses and trapping the maximum amount of solar radiation inside the cooker. This increased solar flux concentration leads to higher absorber temperatures, thereby improving the solar cooker's thermal efficiency.

In this simulation, various inclination angles were tested for the cooker's internal surfaces. These angles were carefully chosen to preserve the receiver surface dimensions of the conventional model, ensuring a relevant performance comparison. The selected angle values are as follows. The following figures (15) illustrate the simulation results:

Figure 15. Evolution of absorber and indoor air temperature of the rectangular solar cooker with an inclined mirror, without load

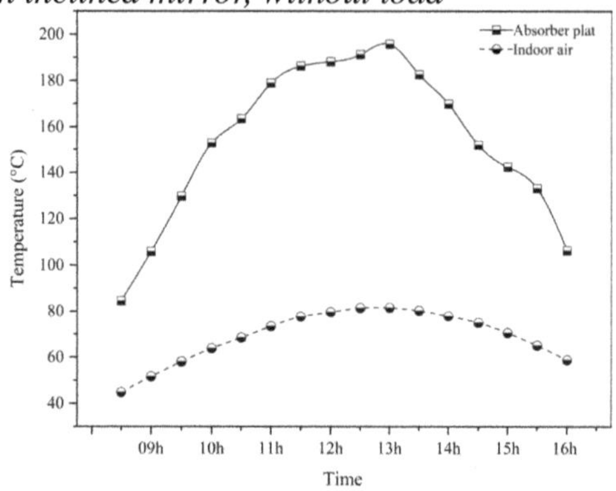

According to the results presented in Figure (15), it is observed that the absorber temperature reaches its maximum value for this case without load, at 195°C for the absorber plate and 95°C for the interior air temperature of the chamber at 13:00. The first figure of merit in this case reaches a value of 0.157 °C·m²/W. This value is higher than the other cases of rectangular, trapezoidal, and triangular shapes, both without load and with single and double glazing. This can be explained by the effect of the internal mirror tilt on the thermal efficiency of the solar cooker.

As with the other cases, a quantity of water load was placed inside the solar cooker with tilted mirrors, and the simulation was rerun to observe the effect of mirror inclination on the concentration of solar radiation and the improvement of the thermal performance of this type of solar cooker. The following figure illustrates the simulation results with the load.

Figure 16. Evolution of absorber plate and water temperature for a rectangular solar cooker with internal mirror inclination with load

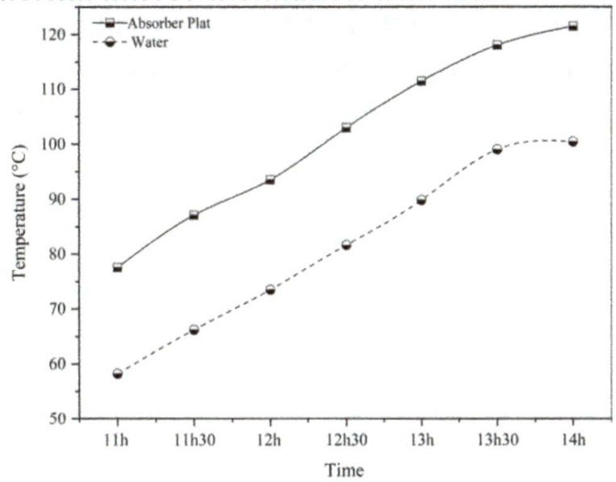

The following figure shows the evolution of the absorber temperature and the water load temperature inside the solar cooker. It is shown in this figure that the maximum water temperature for the rectangular solar cooker with tilted internal mirrors reaches 99°C at 13:00. In this case, after applying the relation (18) to calculate the second figure of merit, it reaches a value of 0.8789. This improvement in the second figure of merit is due to the reduction in cooking time in the simulation and the enhancement of the first figure of merit, F1. The following figures (17) present the distribution of isothermal temperatures within the solar cooker for both cases, with and without load.

Figure 17. Temperature contours for both cases: (a) without load (b) with load, for the rectangular case with inclined internal mirrors

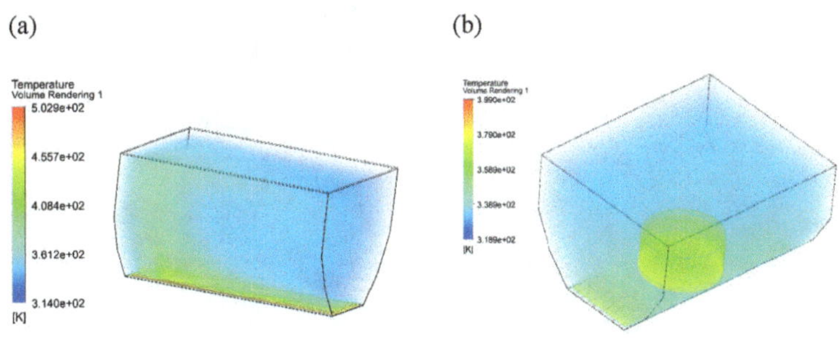

Based on the calculation of the first two figures of merit and the boiling time required for the water load placed in the different geometries presented in this work, and to properly analyze the thermal performance of the solar cooker, it is necessary to calculate the heat loss coefficient created during the heat exchange process between the cooker components and the external environment. Using the relation (13)-(14) and the values obtained for the absorber temperature as a function of time, the following figure presents the results of this coefficient calculation for all the cases studied.

Figure 18. Evolution of the heat loss coefficient for the three cookers with single and double glazing: (a) rectangular, (b) trapezoidal, (c) triangular

(a)

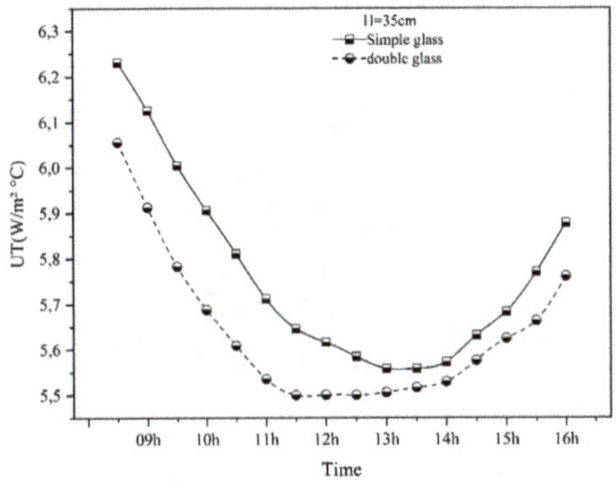

(b)

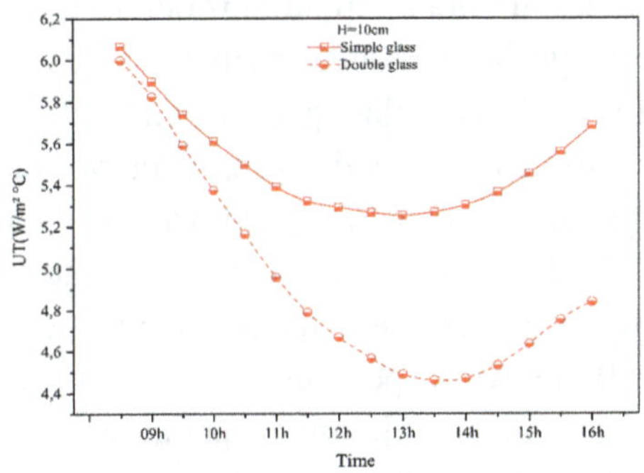

(c)

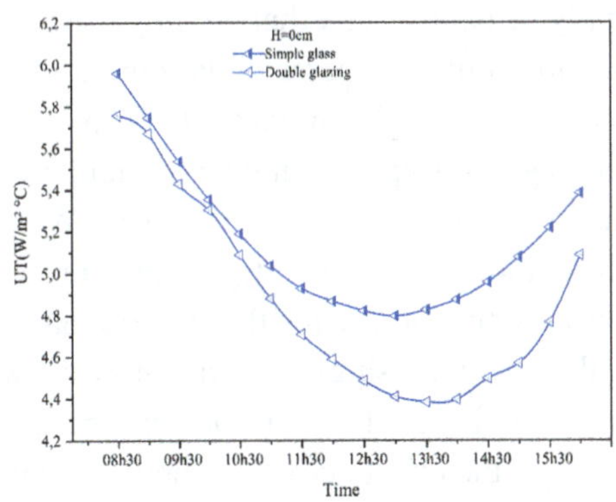

Figure 19. Comparison of the heat loss coefficient for the triangular configuration with double glazing and the rectangular configuration with inclined mirrors

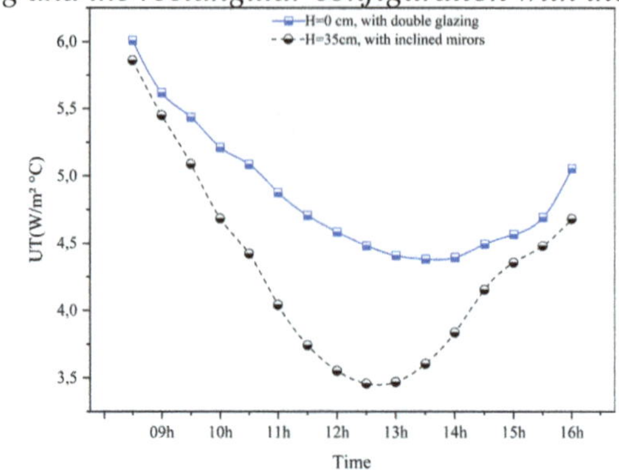

The three presented figures illustrate the evolution of the thermal loss coefficient of solar cookers for two configurations: single and double glazing. It is clearly shown that adding a second glass layer helps reduce thermal losses. This can be explained by the air layer created between the two glass panes, which acts as a thermal insulator, thereby minimizing heat loss. The variations in the curves observed in this study can be attributed to the influence of wind speed, which fluctuates throughout the day. Between 08:00 and 11:30, as well as between 15:00 and 18:00, the wind speed is relatively high, leading to an increase in the thermal loss coefficient. In contrast, between 12:00 and 15:00, the wind speed decreases, reducing thermal losses and allowing the cookers to reach their maximum temperatures. The figures also indicate that decreasing the height of the cooker reduces thermal losses. This is because a lower height enhances the internal temperature of the cooker, thereby minimizing heat loss. Additionally, it is observed that adding a second glass layer further decreases thermal losses in all three types of solar cookers studied. The recorded thermal loss coefficients for single glazing are 5.6 W/m^2. °C, 5.1 W/m^2. °C, and 4.6 W/m^2. °C for the rectangular, trapezoidal, and triangular cookers, respectively. After adding double glazing, these values decrease to 5.2 W/m^2. °C, 4.4 W/m^2. °C, and 4.2 W/m^2.°C for the same cookers. The figure also presents a comparison of the variation in the heat loss coefficient for the two triangular cases with double glazing and the rectangular case with single glazing, this time with inclined internal mirrors. The figure shows that this coefficient is lower for the rectangular case with inclined mirrors, highlighting the influence of geometry in reducing heat losses and improving efficiency.

Global thermal efficiency is a key parameter for evaluating the energy performance of a solar system. In this study, it was calculated for different models of solar cookers.

For a single-glazed configuration, the thermal efficiency values obtained are 7%, 10%, and 16% for the rectangular, trapezoidal, and triangular cookers, respectively. When a second glass layer is added, creating a double-glazed system, these values increase to 10%, 21%, and 31% for the same configurations.

Additionally, in the case of a rectangular solar cooker with inclined internal mirrors, the thermal efficiency reaches 17%. This configuration enhances the thermal performance of the cooker by optimizing solar radiation absorption. As a result, it allows for higher temperatures and reduces cooking time, further improving the efficiency of the system. The observed improvement in thermal efficiency with double glazing is due to the creation of an air gap between the two glass layers, which acts as thermal insulation. This reduction in heat loss helps retain more energy, thereby enhancing the overall performance of the solar cookers.

8 FLAT ABSORBER CONTOURS FOR ALL THE CASES SIMULATED FOR THE SOLAR BOX COOKER WITH ANSYS FLUENT

In this section, the temperature distribution at the absorber surface has been presented for all the cases studied: without load, with single glazing and double glazing, as well as with load for both single and double glazing, for the three cases rectangular, trapezoidal, and triangular. Additionally, the isotherms for the rectangular solar cooker case with tilted internal mirrors will be presented. This will provide the temperature distribution within the solar cookers. All the isotherms presented are for the absorber surface of the solar cookers and the amount of load placed inside the cookers for the different cases studied, at 13:00, when the absorber temperature reaches its maximum. It is also shown in the various curves from the simulation results that the temperature at the absorber surface is highest between 11:00 and 13:00. This explains that this is the optimal time for the solar cooker, as it absorbs the maximum amount of energy, which is then converted into heat.

Figure 20. Temperature contours of the absorber without load and with a single glass cover for (a) the rectangular configuration and (b) for the trapezoidal configuration

(a) (b)

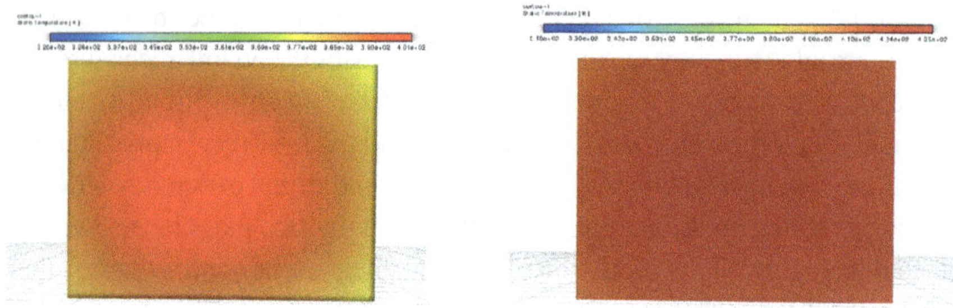

The figure (20) shows the temperature distribution at the absorber surface for trapezoidal and triangular configuration without load and with single glazing. It is observed in the isotherms that the entire absorber surface reaches the maximum temperature, which means that different containers can be placed at various positions on the solar cooker's surface.

Figure 21. Contours temperature for (a) rectangular configuration, (b) trapezoidal geometry and (c) for triangular configuration with double glazing without load

(a)

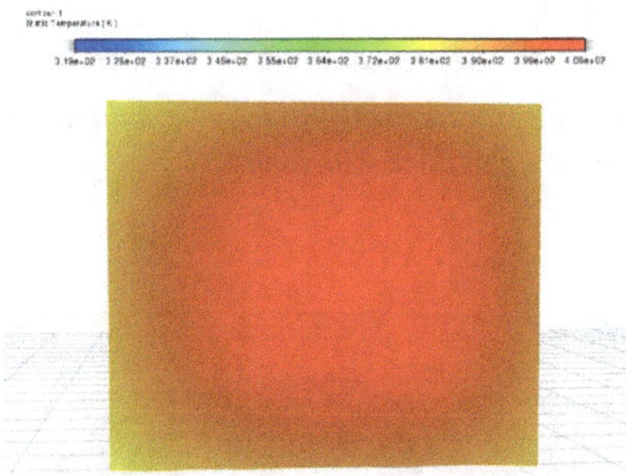

(b)

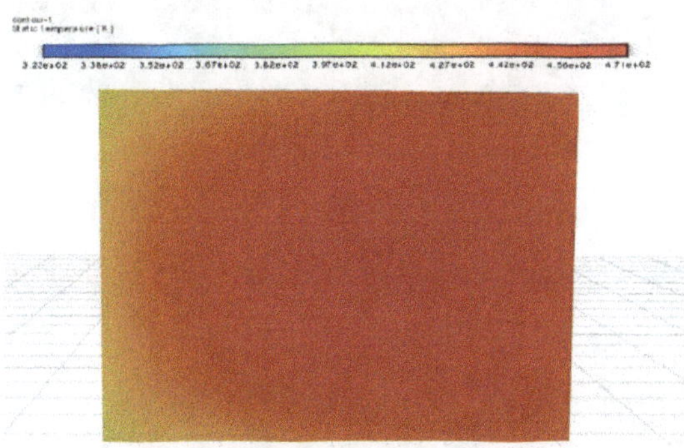

(c)

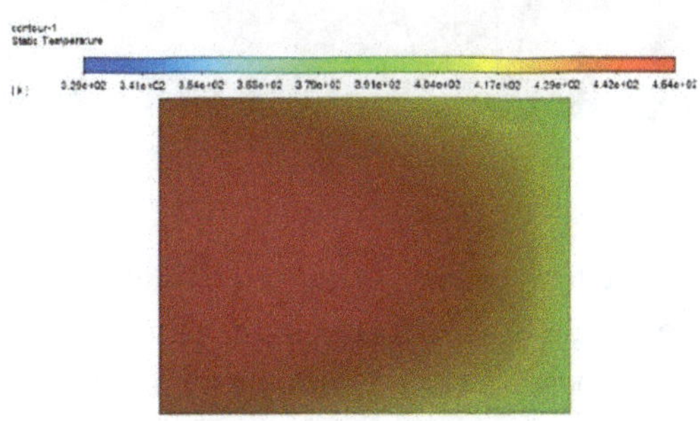

The following figure (21) shows the isotherms of the absorber surface for the three solar cookers in the case with double glazing and no load. It is evident from the presented isotherms that the absorber surface is warmer compared to the case with single glazing, due to the increase in temperature when using double glazing and the reduction in heat losses.

Figure 22. Contours temperature for (a) triangular configuration, (b) trapezoidal configuration with simple glass and with load

(a)

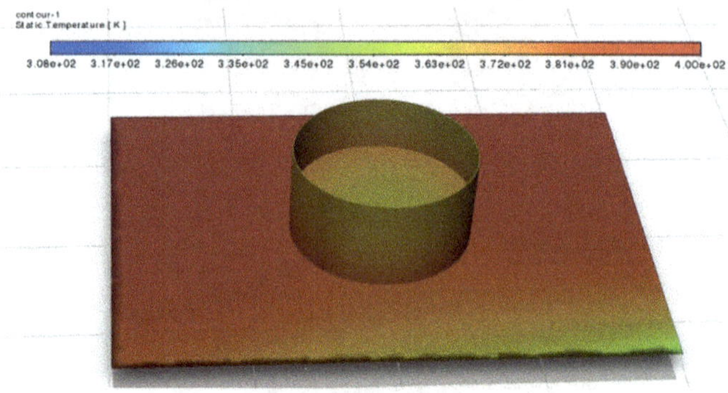

(b)

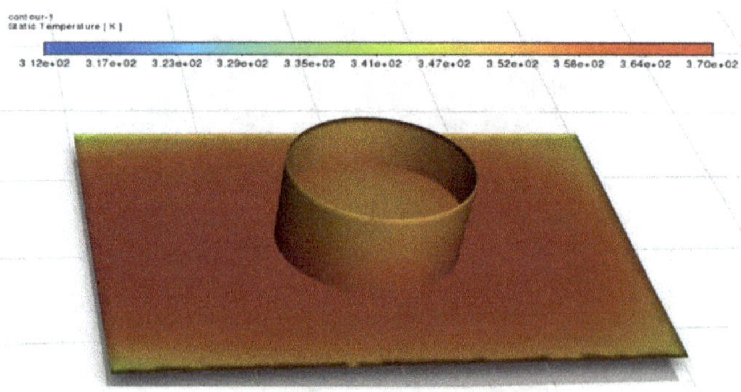

Figure (22) shows the temperature distribution on the absorber surface and the amount of load placed in the solar cookers in the case of single glazing. It clearly illustrates that the load reaches its maximum temperature at this hour.

Figure 23. Contours temperature for water and cylinder and Absorber plat (a) trapezoidal configuration, (c)triangular configuration with double glazing

(a)

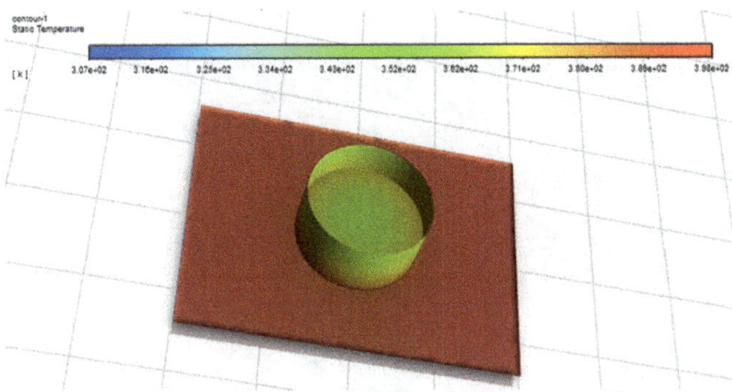

(b)

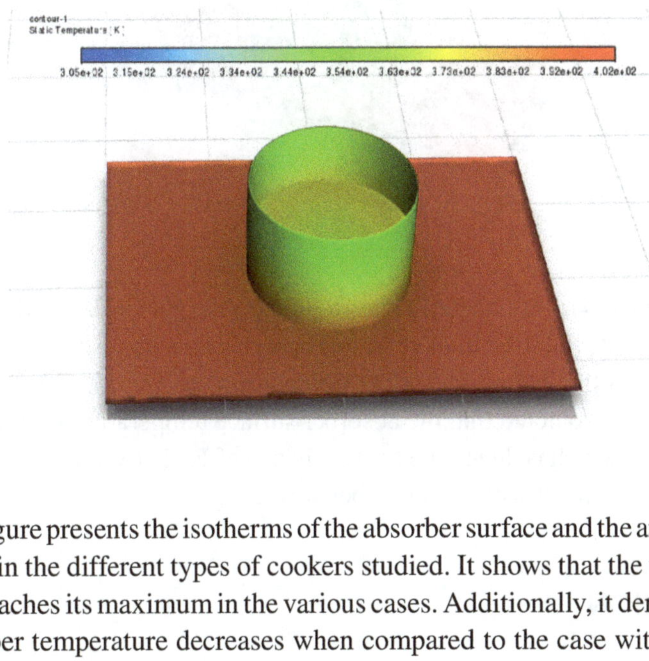

This figure presents the isotherms of the absorber surface and the amount of water load used in the different types of cookers studied. It shows that the temperature of the load reaches its maximum in the various cases. Additionally, it demonstrates that the absorber temperature decreases when compared to the case without load, due to the heat exchange between the absorber surface and the surface of the cylinder containing the load.

Figure 24. Contours temperature for rectangular configuration with inclined mirrors (a) without load, (b) with load

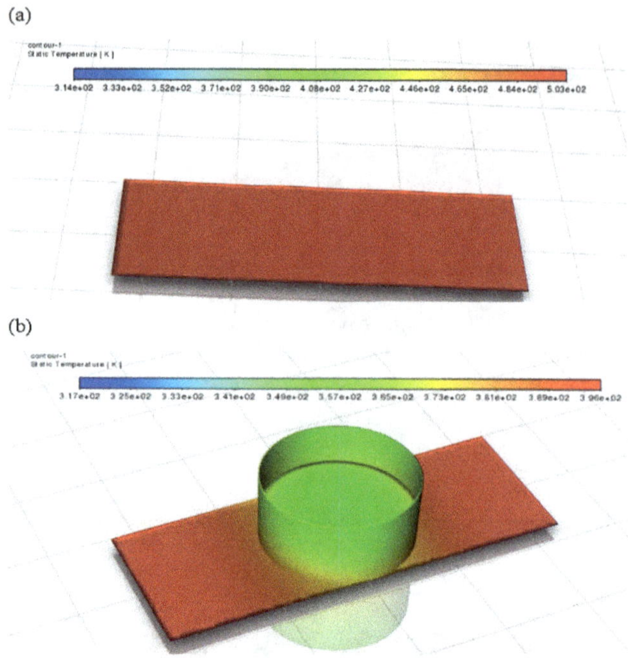

The last figure (24) presents the isotherms of the absorber surface and the amount of load placed inside the solar cooker. In this case, it corresponds to the rectangular configuration with the inclination of the internal mirrors for both cases, with and without load. It is also noted that the absorber surface temperature decreases when the load is added, which is due to the heat exchange effect between the two surfaces of the cylinder in contact with the absorber surface.

9 CONCLUSION

In this study, a thermal performance analysis of different types of solar cookers was conducted using a CFD simulation. The objective was to examine the temperature evolution of the absorber surface under various conditions: with and without load, with single and double glazing, and for different cooker configurations. These configurations varied in height: Hp=35 cm for the rectangular geometry, Hp=10 cm for the trapezoidal configuration, and Hp=0 cm for the triangular case. Additionally, efforts were made to enhance the efficiency of the rectangular configuration using two methods: first, by adding a primary reflector, and second, by tilting the internal mirrors at different angles relative to the horizontal while maintaining the same receiver surface area as the conventional design.

The results clearly demonstrate the impact of cooker geometry on solar radiation concentration and the temperature improvement at the absorber surface across different cases. Moreover, the addition of double glazing significantly enhances the thermal performance of solar cookers by reducing cooking time, optimizing the load capacity, and increasing the absorber temperature. This effect is attributed to the air layer formed between the two glass panes, which acts as thermal insulation, thereby reducing heat losses. This conclusion is supported by the thermal loss coefficient (UT) calculations presented in the results.

Several parameters were calculated to evaluate the thermal efficiency of the solar cookers, including the first and second figures of merit, which are crucial for classifying solar cookers and assessing their heat transfer efficiency from the absorber surface to the cooking chamber. The main findings are summarized as follows:

- Without load and with single glazing, the first figure of merit reached 0.08, 0.099, and 0.1246°C m^2/W, with absorber stagnation temperatures of 112°C, 135°C, and 160°C for the rectangular, trapezoidal, and triangular cookers, respectively.
- Without load and with double glazing, the first figure of merit increased to 0.083, 0.1349, and 0.1452°C m^2/W, with absorber stagnation temperatures of 120°C, 170°C, and 180°C for the rectangular, trapezoidal, and triangular configurations, respectively.
- With load and single glazing, the second figure of merit reached 0.1872, 0.2150, and 0.3609, for the rectangular, trapezoidal, and triangular cases, respectively.
- With load and double glazing, the second figure of merit further improved to 0.3025, 0.5211, and 0.675, for the rectangular, trapezoidal, and triangular configurations, respectively.
- For the rectangular configuration with a primary reflector, the first figure of merit was 0.11°C m^2/W, with an absorber stagnation temperature of 140°C.
- In the rectangular configuration with tilted internal mirrors, the first figure of merit increased to 0.157°C m^2/W.

To further analyze the performance of these different solar cooker designs, calculating the thermal loss coefficient (UT) is essential to evaluate the effect of these modifications on heat loss reduction and overall efficiency improvement. The UT values obtained are:

- With single glazing: 5.6 W/m^2.°C, 5.1 W/m^2.°C, and 4.6 W/m^2.°C for the rectangular, trapezoidal, and triangular cookers, respectively.
- With double glazing: These values decrease to 5 W/m^2 °C, 4.4 W/m^2 °C, and 4.2 W/m^2 °C, respectively.
- For the rectangular cooker with tilted internal mirrors, the UT value further decreases to approximately 3.3 W/m^2.°C, indicating a significant reduction in thermal losses.

This study highlights the importance of geometry, glazing, and internal reflectors in optimizing the thermal performance of solar cookers.

REFERENCES

Abhishek, S. A. X. E. N. A., & Mehmet, K. A. R. A. K. I. L. C. I. K. (2017). Performance evaluation of a solar cooker with low-cost heat storage material. *International Journal of Sustainable and Green Energy*, *6*(4), 57–63.

Abhishek, S. A. X. E. N. A., & Nitin, A. G. A. R. W. A. L. (2018). Performance characteristics of a new hybrid solar cooker with air duct. *Solar Energy*, *159*, 628–637.

ADEME. Zeleke & HAMEER, Sameer. (2018). Design, construction and performance evaluation of a Box type solar cooker with a glazing wiper mechanism. AIMS energy vol. 6, no 1

Adewumi, S. D., Abdulrahim, A. T., Adeleke, A. A., Ikubanni, P. P., & Agboola, O. O. (2019). Performance evaluation of solar cooker made from locally sourced materials. [IJMPERD]. *International Journal of Mechanical and Production Engineering Research and Development*, *9*(2), 299–310.

AKOY, Elamin OM & AHMED, Abdalla IA. (2015). Design, construction and performance evaluation of solar cookers. *Journal of Agricultural Science and Engineering*, *1*(2), 75–82.

AL-NEHARI, Hamoud A., MOHAMMED, Mahmoud A., ODHAH, Abdulkarem A., K.A. Al-attab., Bakal K. Mohammed, Abdulwahab M. Al-Habari, Nasr H. Al-Fahd. (2021). Experimental and numerical analysis of tiltable box-type solar cooker with tracking mechanism. *Renewable Energy*, *180*, 954–965.

AMER, Emad H. (2003). Theoretical and experimental assessment of a double exposure solar cooker. *Energy Conversion and Management*, *44*(16), 2651–2663.

Attou, Y. (2019). Numerical and Thermal Study of a Turbulent Flow in the Annular Space of a Rotor-Stator. Doctoral dissertation, University of Oran 2, Mohamed Ben Ahmed.

Chang, Z. H. O. U., Yinfeng, W. A. N. G., & Jing, L. I.. (2023). Simulation and economic analysis of an innovative indoor solar cooking system with energy storage. *Solar Energy*, *263*, 111816.

Channiwala, S. A., & Doshi, N. I. (1989). Heat loss coefficients for box-type solar cookers. *Solar Energy*, *42*(6), 495–501.

Duffie, J. A., & Beckman, W. A. (2013). *Solar Engineering of Thermal Processes* (4th ed.). John Wiley & Sons Ltd.

Ekechukwu, O. V., & Ugwuoke, N. T. (2003). Design and measured performance of a plane reflector augmented box-type solar-energy cooker. *Renewable Energy, 28*(12), 1935–1952.

El-Sebaii, A. A., & Ibrahim, A. (2005). Experimental testing of a box-type solar cooker using the standard procedure of cooking power. *Renewable Energy*, *30*(12), 1861–1871.

El-Sebaii, A. A., & Ibrahim, A. A. E. S. (2005). Experimental testing of a box-type solar cooker using the standard procedure of cooking power. *Renewable Energy, 30*(12), 1861–1871.

Harmim, A., Merzouk, M., Boukar, M., & Amar, M. (2012). Performance study of a box-type solar cooker employing an asymmetric compound parabolic concentrator. *Energy*, *47*(1), 471–480.

IBRAHIM, Said MA., & EL-REIDY, Medhat K. (2007). The performance of a solar cooker in Egypt. *Renewable Energy*, *6*(8), 1041–1050.

ISHAQ, Muhammad., OLUWOLE. Fasiu Ajani., AMULAH. Nuhu Caleb., Bello Abdulsalam. (2023). Performance Evaluation of Box-Type Solar Cookers Using Different Insulation Materials. Nigerian Journal of Science and Engineering Infrastructure vol. 1, no 1.

Jones, W. Peter, LAUNDER, Brian Edward. (1972). The prediction of laminarization with a two-equation model of turbulence. *International Journal of Heat and Mass Transfer*, *15*(2), 301–314.

Jubran, B. A., & Alsaad, M. A. (1991). Parametric study of a box-type solar cooker. *Energy Conversion and Management*, *32*(3), 223–234.

Jubran, B. A., & Alsaad, M. A. (1991). Parametric study of a box-type solar cooker. *Energy Conversion and Management*, *32*(3), 223–234.

KHALDI. Souheyla Yahiaoui. (2018). Numerical Study of the Thermal Behavior of a Solar Dryer Using a Thermal Bed for Energy Storage (2018). Doctoral dissertation, University of Bourgogne Franche-Comté; Abou Bekr Belkaid University (Tlemcen, Algeria).

KULDHAR. Santosh & SHRIVASTAVA, Ritu. (2023). Development and Fabrication of Box Type Solar cooker with PCM to Enhance Its Efficiency, vol 10.

KUMAR, Naveen, AGRAVAT, Sagar, CHAVDA, H.N. Mistry. (2008). Design and development of efficient multipurpose domestic solar cookers/dryers. *Renewable Energy*, *33*(10), 2207–2211.

Mirdha, U. S., & Dhariwal, S. R. (2008). Design optimization of solar cooker. *Renewable Energy, 33*(3), 530–544.

Nahar, N. M. (1990). Performance and testing of an improved hot box solar cooker. *Energy Conversion and Management, 30*(1), 9–16.

Nahar, N. M., Marshall, R. H., & Brinkworth, B. J. (1993). Studies on a hot box solar cooker with transparent insulation materials. *Energy Conversion and Management, 35*(9), 787–791.

Nahar, N. M., Marshall, R. H., & Brinkworth, B. J. (1994). Studies on a hot box solar cooker with transparent insulation materials. *Energy Conversion and Management, 35*(9), 787–791.

SAXENA, Abhishek, CUCE, Erdem, TIWARI, G. N., Avnish Kumar. (2020). Design and thermal performance investigation of a box cooker with flexible solar collector tubes: Experimental research. Energy, vol. 206, p. 118144.

ANILKUMAR, B. C., MANIYERI, Ranjith, et ANISH, S. (2021). Optimum selection of phase change material for solar box cooker integrated with thermal energy storage unit using multi-criteria decision-making technique. *Journal of Energy Storage, 40*, 102807.

Swinbank, W. C. (1963). Long-Wave Radiation from Clear Skies. *Quarterly Journal of the Royal Meteorological Society, 89*, 339–348.

Uhuegbu, C. C. (2011). Design and Construction of a Wooden Solar Box Cooker with performance and efficiency test. *Journal of Basic and Applied Scientific Research, 1*(7), 533–538.

WASSIE. Hailemariam M., GETIE, Muluken Z., ALEM, Mulat S., *Teshome B., Zelalem M., 2022).* Experimental investigation of the effect of reflectors on thermal performance of box type solar cooker. *Heliyon* vol. 8, no 12.

WASSIE. Hailemariam M., GETIE, Muluken Z., ALEM, Mulat S., Kotu Teshome B., (2022). Experimental investigation of the effect of reflectors on thermal performance of box type solar cooker. Heliyon vol. 8, no 12

Weldu, Amanuel, ZHAO, Li, DENG, Shuai, Mulugeta. Nigussie, Zhang. Ying, Nie. Xianhue, Xu. Weicong. (2019). Performance evaluation on solar box cooker with reflector tracking at optimal angle under Bahir Dar climate. *Solar Energy, 180*, 664–677.

APPENDIX

Nomenclature

F_1: First factor of merit (°C m^2 /W)
T_P: Stagnation temperature for absorber plat (°C)
T_{am}: Ambient temperature (°C)
F_2: Second factor of merit (dimensionless)
T_{W1}: Initial temperature for water (°C)
T_{W2}: Finanl temperature for water (°C)
Δ_T: is temperature difference between maximum temperature of water and the ambient temperature (°C)
Δ_t: is time required to achieve the maximum temperature of cooking
A: area of aperture of the cooker (m^2)
M_W: mass of water (K_g)
C_W: specific heat capacity of water J K_g^{-1} K^{-1}
C_μ: Constant for the k-ε turbulence model
K: Turbulence kinetic energy, m^2 s-^{2}
T_{SKY}: Sky temperature, K
UT:
N: Number of glasses
h_0: Convective heat transfer coefficient Wm^{-2} K^{-1}
V_{wind}: Wind velocity m s^{-1}
ε_p: Emissivity of absorber plat
ε_g: Emissivity of glass
β: Angle of inclination of the glass surface
σ: The Stefan-Boltzmann constant
CFD: Computational Fluid Dynamics
T_{Sky}: sky temperature
e_g: Glass thickness
e_p: Absorber thickness
e_{iso}: Insulation thickness
e_w: Wooden box thickness
φ': The optimal tilt angle of the main added reflector for solar cookers

Chapter 6
Solar Clean Cooking System With the Aid of Power Converters:
Towards Sustainable Cooking Revolution

S. Muthukaruppasamy
https://orcid.org/0000-0002-7978-4236
Velammal Institute of Technology, India

Thomas G. Arun Sampaul
https://orcid.org/0000-0002-5438-5781
J.B. Institute of Engineering and Technology, India

J. Nandha Gopal
https://orcid.org/0000-0003-3301-484X
Velammal Institute of Technology, India

E. Parimalasundar
Mohan Babu University, India

ABSTRACT

Energy conservation has been a primary focus for scientists. A significant issue of the present is the fast depletion of earth's resources. It is important to diminish reliance on non-renewable resources. Efforts in solar cooking have consistently shown their advantages and efficacy. Solar cooking is a technology that derives energy from photovoltaic solar panels driven by sunshine. We possess a technique that stores solar energy in a battery pack for use during periods without sunlight.

DOI: 10.4018/979-8-3373-2535-4.ch006

Electrical energy is generated by photovoltaic panels (PV) and stored in a battery. Consequently, the cookers are energized by boost converter. The integration of photovoltaic panels and batteries will provide the necessary electricity during daylight hours. A comprehensive primary grid backup is also included into the system. This article investigates electric cooking system aimed at advancing renewable and sustainable cooking technology with the application of power electronic converters

1.INTRODUCTION

1.1. Why is Cooking in Solar PV?

Existing clean cooking technologies fulfil the fundamental requirements of the most disadvantaged people impacted by humanitarian emergencies. Significant changes in the choice of culinary solutions and the mobilization of funds are necessary to facilitate large-scale migrations away from biomass. E-cooking is an essential technology that may facilitate a carbon-free energy transition, particularly when fuelled by renewable electricity, whether sourced from the grid, micro-grids or independent off-grid power sources, such as solar photovoltaic (PV) systems (Teske et al., 2023). E-cooking employs electric cookers, induction stove tops, and several other electric cooking devices. This method presents numerous benefits compared to conventional cooking techniques, such as enhanced quality of indoor air, minimized deforestation, improved efficiency of energy, and the capacity to utilize renewable energy sources to mitigate the effects of greenhouse gas emission levels. The application of solar power has emerged as a particularly promising resource. This is particularly important for underdeveloped countries with restricted access to power. Highly environmentally friendly e-cooking equipment, including electric pressure cookers, rice cookers, and induction cookers, consume significantly less power than inefficient alternatives, such as hot plates, utilizing only 2-4 kWh per day compared to 10-30 kWh (Scott et al., 2024). This renders them not only feasible but also economically competitive with other cooking methods, provided that electricity is available at reasonable rates. Furthermore, if power is sourced directly from a solar photovoltaic system, eliminating continuous energy expenses, this solution emerges as the most economical alternative currently accessible (Panagoda et al., 2023). In the domain of photovoltaic e-cooking devices, they referenced the advancement of photovoltaic-powered electric culinary, alongside storage solutions including battery storage, phase change materials (PCM), and insulated cookers. In the domain of interconnection with mini- and micro-grids, many global initiatives that include photovoltaic cooking into dispersed grids and the potential contributions of machine learning predictions. Regarding hydrogen-based cooking using photovoltaic technology, it was emphasized that it may achieve cost competitiveness by 2030; nevertheless, difficulties related to storage, infrastructure, and safety remain unresolved (Mukelabai et al., 2022). Research on socio-economic and adoption issues indicates that behavioural findings reveal the effect of many factors on adoption rates (Dessart et al., 2019). Prospective studies should concentrate on the integration of artificial intelligence and the Internet of Things (IoT) for predictive framework optimization, the development of creative thermal and blended storage methods, the exploration of adaptable hydrogen-based cooking methods, the execution of economic and social analyses of user actions, and the evaluation of both ecological and fiscal effects through life-cycle assessments.

1.2. Solar PV E-Cooking

Photovoltaic e-cooking utilizes sun energy captured by solar panels to operate contemporary cooking devices. The core system consists of solar photovoltaic panels that produce direct current power, usually managed by a charge controller. It could involve an inverter to change direct current conversion to alternating current for traditional appliances. Figure.1 illustrates a suggested photovoltaic e-cooking system.

Figure 1. PV-eCooking

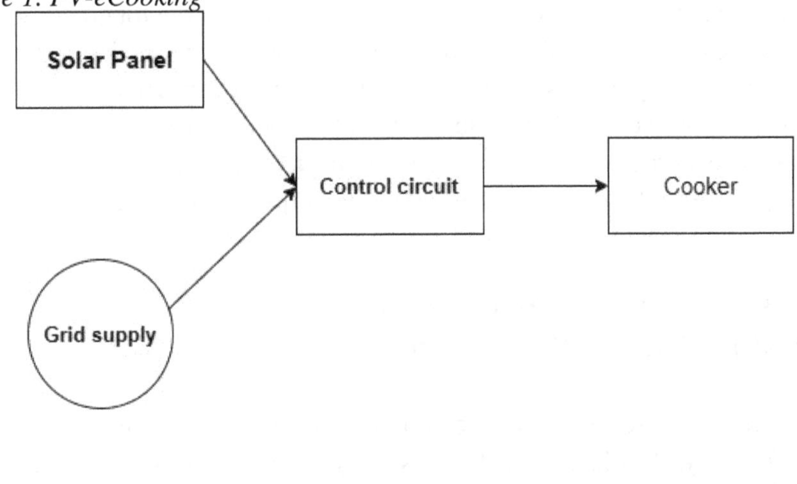

There are various configurations for sunlight PV-based e-cooking platforms. The easiest and affordable option is direct solar power cooking, which only allows cooking during the daytime. Another configuration, battery-integrated, allows cooking at all times through storage of energy. Finally, a hybrid setup combines solar PV with power from the grid for highest reliability. Also, it may work in tandem with substances that store thermal energy, such as tangible heat or hidden warmth. Usually energy-saving devices especially made to run efficiently inside the power limits to solar panel, this mechanism is becoming more and more relevant for clean cooking projects all around. Solar PV electric culinary is developing to three main categories: Evolution in solar PV electric culinary gadgets; amalgamation of PV electric cooking to micro grids; solar PV-hydrogen based cooking and energy availability, acceptability, and socioeconomic difficulties.

The preface of the article is carried out on the following objectives:

i)The objective of this article is to investigate a comprehensive examination of research on electric cooking system aimed at advancing renewable and sustainable cooking technology with the application of power electronic converters with real world examples

ii) The objective is to explore the effective advancements in solar PV-powered culinary equipment and addition of PV e-cooking with micro grid with inclusion of green hydrogen cooking.

The remaining part of this article is focused as follows: Section.2 provides an innovation in solar photovoltaic technology for cooking and outlines the incorporation of PV e-cooking with micro grid. In addition, section 2 describes renewable hydrogen cooking and socio-economic cultural consideration is discussed for solar electric cooking. Section 3 discusses role of MPPT and converter for green cooking. Sections 4 focus the description of converter and inverter and section 5 discusses the real world examples of solar PV e-cooking. Section 6 provides the potential study prospects for the e-cooking industry. Finally, conclusion part concludes recent advances and patterns in solar PV-based e-cooking technologies, showing how they could help green cooking issues, especially in areas that lack electricity availability.

2. ADVANCEMENTS IN SOLAR PV-POWERED CULINARY EQUIPMENT

This part emphasizes how innovations in solar photovoltaic technology are enhancing the efficiency, affordability, and accessibility of e-cooking equipment, therefore fostering equitable growth and energy safety. The feasibility of incorporating thermal power retention into off-grid photovoltaic systems is assessed to diminish cooking consumption of energy (Paneru et al., 2024)]. The methods typically categorize into two types: sensible heat storage and latent heat storage, utilizing water, oil, and phase change materials. The accumulated thermal power can substantially diminish the amount of electrical power required for cooking, particularly at periods of high demand. In solar photovoltaic systems with sensible heat storage, solar PV provides electricity to heat substances such as water, oil, or rocks, which retain thermal energy for cooking at non-sunlight time. Essential attributes include economical and uncomplicated, reduced energy density, efficiency contingent upon temperature. For example, in (Chiloane et al., 2023], a laboratory experiment shown that pre-heated water at temperatures ranging from 51 to 96.6 oC can significantly reduce energy consumption in e-cooking. Energy savings for cooking rice varied between 11.36% and 28.79%. This reduction also results in diminished battery capacity requirements for off-grid photovoltaic systems. Techniques for pre-heating water utilizing thermal energy retention for culinary purposes encompass systems that absorb and retain heat from renewable or waste energy sources for subsequent utilization. In solar photovoltaic systems with latent heat storage, solar PV generates energy to charge phase change materials that retain thermal energy via state transitions for utilization at non-sunlight time. Essential attributes are Elevated energy density, approximately stable temperature throughout phase transition, costlier than sensible heat storage, comprising components such as paraffins and salt hydrates.

Further, modest type solar box stove is transformed into a photovoltaic with solar thermal stove to enhance effectiveness of cooking and user convenience (Joshi & Jani, 2015; Misra et al., 2023). It demonstrates enhanced efficiency and decreased cooking duration. The cost factors suggest possible affordability and appeal with more commercialization. Photovoltaic -powered induction cooking stoves have been constructed, revealing that the use patterns of these stoves provide essential insights into their efficacy and community acceptability (Damayanti et al., 2023). The streamed solar electric cooker has been created as a neat, odourless culinary method. It employs diodes as heating components and phase change materials for heat storage, exhibiting effective culinary skills with rapid durations of culinary, as seen in Figure.2

Figure 2. Streamed solar electric cooker with phase changing material

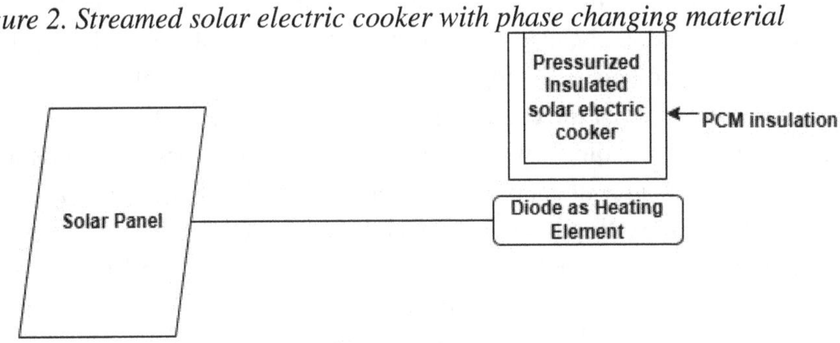

From the findings, photovoltaic system electric cooking can power the phase changing material and maintain moderate temperature for cooking rice. It facilitates a heating duration of 60 minutes when completely get charged, offering a neat and effective option. Solar-powered insulated cooker is created using erythritol phase change material (PCM) for energy storage (Felix et al., 2025). Solar energy system(110KW) can gently heat five kilos of food overnight. Moreover, erythritol PCM with two kilograms may retain about 0.3 kWhr power, facilitating rapid culinary. Moreover, the heat retention capability is comparable to that of pricier systems using battery storage and stove tops. An economical and competent solar cooker has been created, including electrical backup and phase change material energy repository to guarantee uninterrupted cooking at non-sunlight hours. Coconut oil had the greatest efficacy evaluated amid phase changing material, with thermal efficiency of 0.139. By preserving coconut oil, the suggested solar culinary system achieves an energy level of 39% with backup power. Solar panel is connected by solar electric cooking which is linked to an electric heater situated with insulated bed chamber (Batchelor et al., 2018). SEC technologies, when used with suitable insulation, exhibit cost competitiveness. Initial findings indicate the feasibility and efficacy of solar culinary in delivering effective and clean cooking result. A photovoltaic-charged electric double-burner stove successfully mitigates energy shortages. Favourable outcomes are achieved for cooking duration for many dishes, demonstrating the stove's suitability for domestic use. A affordable green cooking option is achieved by integrating solar photovoltaic technology with the electric grid, therefore obviating necessity to grid connected inverter and using a novel power saving approach. Expenditure of energy using this method for culinary is small at 4.7 USD/month, signifying decrease of 31 percent relative to conventional grid-tied cooking method. Moreover, eliminating grid-tied inverter reduces expenses of energy and streamlines system mechanism. The steel pots of certain size demonstrated the greatest energy transmission and temperature elevation. The PV-IC attained energy efficiency of up to 47.6% at certain intervals. Cooking times with the PV-IC exceeded those of LPG burners but were shorter than those of electric cookers. Additionally, using the PV-IC incurs a cost of 3.88 USD for daily working. Solar photovoltaic technology is coupled to induction cooker via inverter to improve the soundness of the cooking process (Ali et al., 2025). It observed the use of photovoltaic technology which enhances the framework's dependability through stable, green source of energy. This connection facilitates user management of solar and induction heating devices, enhancing cooking ease. The viability and anticipated advantages of e-cooking are assessed with solar-powered homes and minimum-power culinary appliances (Saxena et al., 2022). The research examined the regional area, circumstance effects, and financial expenditures of cooking in comparison to conventional techniques. Larger families need more substantial photovoltaic units in high glaze regions for regularly fulfill e-cooking needs. Moreover, electric cooking with solar modest systems decreases greenhouse gas emissions relative to conventional techniques, leading to a reduced environmental impact. Furthermore, the expense associated with electric cooking is less to LPG. The solar-powered electric cook stove is designed to combine off line grid photovoltaic unit with a minimum energy electric cooking device, therefore overcoming constraints of conventional thermal plus solar cooking systems. The photovoltaic system is expected to provide electricity to additional low-demand domestic electrical loads, hence diminishing dependence on conventional biomass for cooking fuel, fostering sustainable food production, and mitigating environmental pollution. The feasibility of battery-powered electric cooking is examined as a revolutionary approach to provide economical, dependable, and sustainable energy for culinary purposes, leveraging breakthroughs in electrification and the declining prices of solar photovoltaic systems and batteries made of lithium-ion (Leary et al., 2021). This method may be economically viable compared to existing cooking fuel costs, and families mostly preserved their meal selections after the switch. The research indicated that energy-efficient appliances enabled cooking with a daily use of just 0.87–2.06 kWh per family. The cost-controlled solar off line grid photovoltaic cooking compared to wood-based cooking is examined in poor income homes. The research evaluated off line-grid solar culinary methods: direct solar dc power operated through specialized cook stove and energy kept in lithium batteries, with prices varied according to use current market. Their optimum levelled cost of energy research revealed that direct solar dc power can compete with fire wood culinary. The converging cost models imply comparable price-effectiveness and possibilities for further development. A solar PV module-based inverter is modelled for residential cooking applications. The research deployed to model solar panels in diverse configurations and utilized a maximum power point tracking (MPPT) technique to enhance power extraction under varying ecological circumstances. The investigators assessed the electrical electric current, voltage, and energy output of several solar modules at different temperatures and irradiation conditions, validating simulation findings against conceptual estimates for uniformity and precision. It shows and analyzes moderate power photovoltaic unit system for residential culinary, addressing DC and AC loads. The outcomes matched the fulfilment of sample photovoltaic modules and providing an accessible platform for comprehending and analyzing various kinds of photovoltaic modules and their attributes (Ansari et al., 2021). The solar cooker consists photovoltaic panels, DC/DC converters, thermal components, and charging controller that regulates the functions according to customer specifications. The research demonstrates that every day's energy generation may be accumulated in solar batteries, facilitating culinary during daylight and night-time. The results illustrate the feasibility of the suggested autonomous solar cooker for both urban and rural families, offering an ecologically sound cooking alternative, particularly in regions with restricted or intermittent energy supply. Solar electric cooking prototypes are created to directly interface with solar panels for culinary and domestic electricity purposes (Redekal & Deb, 2024). This model needs a minimal expenditure, creating them attainable for workers in Hong Kong. The efficacy of direct solar DC power is shown to culinary and electric charging gadgets (Rashid et al., 2024). The study sought to show how reasonably priced solar dc power and ideas like insulating the culinary bed chamber would decrease reliance on conventional culinary fuel. Studies indicated that heaters capture greater amounts of energy from solar cells compared to resistive heaters, hence improving heat generation for cooking. The power cost of Direct DC Solar is comparable with biomass culinary in several places, underscoring its capability for providing inexpensive electric power access. Implementing Direct DC Solar may provide several advantages, like diminished indoor air pollution, less deforestation, and lowered financial obligations related to conventional cooking fuels. Studies have shown that cooking necessitates much less electricity and energy than often presumed, indicating solar culinary as an economic electricity option. Economic study indicates that suggested solar culinary may be financially feasible, especially in offline-grid regions, while included into solar structured household unit. An electric burner driven by photovoltaic energy has been created to the efficient utilization of photovoltaic energy (Prasanna & Umanand, 2011). The engineered stove, fuelled by photovoltaic solar energy, offers a practical remedy to inadequacy of LPG supply for cookery in developing nations. Furthermore, study illustrates economic feasibility of the electric stove in comparison to conventional cooking techniques.

2.1. Adding E-Cooking With PV to Micro Grids

Integrating photovoltaic electric culinary to grid not only enhances energy efficiency and lessens dependence on fossil fuels but also promotes cleaner cooking, reduces air pollution, and cuts energy costs for communities. This approach mandates meticulous planning, including evaluating local energy needs, optimizing solar generation and storage systems, and deploying accessible technologies. Moreover, it facilitates energy access in off-grid regions, legitimizes women by minimizing the time consumed on fuel collection, aids climate change mitigation efforts, making it a holistic solution for sustainable development. The evaluation of adaptable dimension and less economic constraint for rooftop solar PV combined with lithium-ion battery system in grid-connected households involves analyzing energy consumption patterns, costs, and potential savings. All-electric homes typically benefit more from solar PV and BESS due to reduced reliance on fossil fuels, while homes utilizing both gas and electricity may face varied economic dynamics based on their energy use and utility rates. The study likely assesses factors such as installation costs, incentives, energy prices, and the return on investment to determine the most efficient configurations for each type of household, ultimately aiming to identify solutions that enhance sustainability and economic viability (Delapedra Silva et al., 2022). The integration of a solar e-cooking with an off-grid PV system provides a promising solution to reduce traditional gas cooking, thereby progressing cleaner culinary practices and minimizing environmental pollution. Research in Ethiopia indicates that optimizing culinary time to middle of day can lead to reduced costs for mini-grids, primarily due to the high power demands of e-cooking appliances (Leach et al., 2021). However, shifting these hours has only a modest effect on the overall expenses associated with mini-grid component sizing, with only a 3% reduction in total present cost. This highlights the complexities in balancing energy supply and demand while transitioning to renewable energy solutions in residential cooking. In predicting inordinate power to culinary requirements at PV grid arrangement, machine learning algorithms are employed to effectively capture the complexities of energy generation and consumption patterns, providing valuable insights for optimizing energy use in off-grid settings.

2.2. Green Hydrogen Cooking

Solar PV-based hydrogen is implemented for cooking using solar photovoltaic systems to produce electricity. It drives electrolysis, which breaks water onto hydrogen and oxygen as depicted in Figure.3. The produced hydrogen could be kept and utilized as a renewable energy source for culinary uses, offering renewable substitute to oil and gas. This technique not only helps to decrease the greenhouse gases, but it also promotes energy independence, particularly in areas lacking access to traditional energy sources. By integrating solar energy with hydrogen production, it creates a versatile energy solution that aligns with global efforts toward decarbonization and sustainable living. Recent studies indicate that integrating solar power, batteries, and power-to-hydrogen technologies may represent a cost-effective solution for energy storage and operation of mini-grids. This combination allows for efficient energy conversion and storage, enabling the utilization of excess solar energy for hydrogen production, which can then be utilized for cooking and other applications. The investigations indicate that these blended systems can enhance reliability of energy with sustainability in off-grid environments, although the optimal configurations may vary based on specific local conditions and requirements. As depicted in Figure, Hydrogen creation from green sources, like solar, predominantly utilizes electrolysis, where an electrolyser—specifically a proton exchange membrane (PEM) electrolysis cell—facilitates the breaking water into hydrogen and oxygen. This method is favored for its efficiency and ability to directly transform into hydrogen, enabling the storage and transportation of energy. As the technology advances, electrolysis is expected to have a critical role in change to a hydrogen economy, with applications across several domains, notably transit, industrial operations, and storage of energy(Sadeq et al., 2024; Łosiewicz, 2024).

Figure 3. Solar PV-driven hydrogen cooking

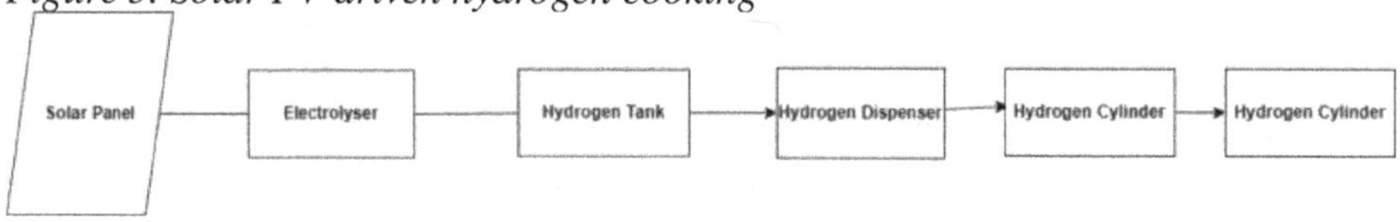

2.3. Social Class Stratum, Economic and Traditional Considerations for Solar PV Electric Cooking

Understanding the financial and cultural aspects that influence energy choices and user behaviors is crucial for favorable reception and long running-term effects of technology solutions, notwithstanding their necessity. This viewpoint is important because it acknowledges that plan and execution of solar PV and culinary services needs to take into account customer preferences, load pattern, and hurdles. In this regard, e-cooking powered by solar photovoltaics has grown into a game-changing alternative for meeting culinary needs of low-income homes (Sarangi et al., 2024). Significant prospects for broad use of Solar-based e-cooking solutions are created by the continuous drop in worldwide PV rates as well as developments in battery science. Current investigations have shown that battery-supported e-cooking could also help national/mini/micro-grids, despite fact that early research concentrated mainly on solar household systems. Investigations counter the conventional belief that high-power energy sources are necessary for cooking. Research underscores that constant temperature regulation, than elevated immediate power, is essential for effective cooking. This revelation demonstrates that cooking may be accomplished with markedly reduced power and energy, so establishing solar cookery as an exceptionally resource-efficient alternative. Price evaluations demonstrate that the incorporation of solar heaters into effectively solar household setups could be cost-effective, especially in off line-grid regions. Furthermore, the implementation and enduring viability of novel renewable energy sources are probable while they correspond with overarching community requirements—such as enhanced kitchen conditions, decreased wood expenses, and superior indoor air quality (Ronge et al., 2016). Ultimately, product features such as resilience, affordability, and measurable enhancements in quality of life are recognized as crucial determinants affecting the adoption and continued utilization of contemporary technology.

3. ROLE OF MPPT AND CONVERTER IN SOLAR PV E-COOKING:

Sunlight can be used to heat your home in two ways: (i) directly and (ii) indirectly. Radiators are utilized for both warmth and making food, and direct warming means that the radiator has to soak up the sun's energy. The solar skillet is the best proof of this. In the indirect way, a PV array turns solar energy to electricity, which is applicable for purpose, like food preparation and heating. It is likely that the indirect approach will become crucial for food preparation and heating, since solar power is getting cheaper every day and more efficient. A significant issue with the solar system is that it doesn't always produce power; it changes based on the environment. It is possible to put a storage device, such as a battery, across the load to meet its steady voltage needs and set the rate of power. The battery does the following:

i)To maintain the level of voltage through the stable load.
ii)Based on how much power the PV system can provide, the battery either soaks up or releases power to maintain the load's power needs stable.

The primary obstacles to the marketing of the culinary PV module are the higher cost of the battery, its upkeep, and environmental concerns during disposal. However, the effectiveness and dependability of the system will not be significantly impacted by variations of power at solar array caused by changes in temperature (T) and irradiance (G) for heating purposes. Therefore, solar electricity can be used without a battery for cooking. However, with the aid of an appropriate MPP tracker, a photovoltaic cell can be run in high power in order to observe the most solar power possible. This necessitates choosing a converter to link the solar panels and thermal load. So that converters duty cycle decides the voltage which must match maximum point voltage and it varies with temperature and radiation. The photovoltaic panel's inner resistance at MPP must match an equivalent load resistance in order to reach the VMPP point and monitor the highest possible power output. Therefore, resistance to load influences the choice of dc-dc converter. In independent mode, the battery is necessary to control the flow of power if the load needs an uninterrupted supply. However, in other applications, such as heating and cooking, the change in load power has little effect on the system's reliability. In those situations, the battery should not be used to avoid the frequent maintenance. To prevent the expense, regular upkeep, and environmental problems brought on by battery consumption, the battery may not be used in following scenarios. Below applications are the usages of solar pv panel with battery in rural regions.

- In a cooking application, a continuous power load is preferable to meet power demand requirements. However, the battery is optional and may or may not be needed.
- In heating application, a continuous power load is preferable to meet power demand requirements. However, the battery is optional and may or may not be needed.
- In Illumination Applications, a continuous power load is absolute necessary. So battery must be required for uninterrupted supply.
- In Motor drive application for irrigation, a continuous power load is absolute necessary. So battery must be required for uninterrupted supply.

3.1 Choice of Converter

The hindering properties of environment create a significant shift in solar irradiation when it reaches the PV array. The peak power point is shifted by changes in irradiation over time. In addition to radiation, temperature changes alter the PV panel's electrical properties. Temperature changes more slowly than radiation does. Therefore, in order to prevent delays and failures in monitoring maximum power, the MPP tracker needs to be constructed as per effective conduct of irradiation. By regulating the duty ratio of converter, MPPT requires that the voltage of PV array remain in MPP under specific irradiance and temperature conditions as shown in Figure.4.

3.1.1. Choice of Converter With Battery

If the battery is integrated across the continuous load, the choice of dc-dc converters, such as buck, boost, or buck-boost converter, is based on battery voltage. If the battery voltage (V_b) is greater than VMPP under the conditions of full irradiance and minimum temperature, the boost converter is utilized to deliver power upon the PV array to the loading. When the battery voltage is lower than VMPP under the conditions of maximum radiation and minimum temperature, the buck converter is employed. Any voltage of the battery could be utilized with a buck-boost converter. Additionally, the capacity of the battery to charge and discharge is dependent on the PV panel's maximum electrical power consumption.

Figure 4. MPPT controlled solar PV with converter

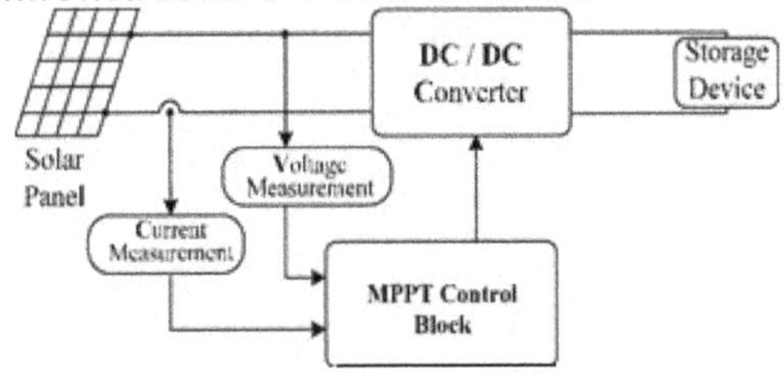

3.1.2. Choice of Converter Without Battery

The theory of maximum power transfer is used to choose a dc-dc converter without a battery across the load. The equivalent resistance of the load applied to the dc-dc converter's input lines must match the internal resistance of the PV panel at MPP for a specific temperature and irradiation in order to maximize power extraction from the panel. For MPPT control, it is concluded that while boost converters are employed while the input resistance of panel is lesser than load resistance, buck converters are utilized while the input resistance exceeds than load resistance to achieve more power deliver from PV panel to final load. Buck-boost converters are independent of load resistance and are used to regulate situations in which the equivalent load resistance fluctuates between 0 and infinity.

3.1.3. Noise Mitigation in Solar PV Cooker Circuits

Electrical circuit noise is a consideration when selecting a converter for a solar PV cooker; specifically, the converter produces electromagnetic or audible noise due to its switching components, necessitating the choice of a converter engineered for quiet functioning or the implementation of noise cancellation filters to reduce interference. The sources of noise in solar photovoltaic cooker circuits are as follows.

i)Electronic and Electromagnetic Interference

Digital components in charge controllers and inverters can generate electrical noise, potentially resulting in radio-frequency interference or non-radiated electro-magnetic interference.

ii)Switching Noise

Numerous charge controllers deliver power to batteries in pulses, and inverters, particularly those employing high-frequency switching, may generate a discernible audible hum or buzz.

iii)How to Mitigate Noise in Solar PV Cooker Circuits?

a) Selection of appropriate converter type

Suitable converters are engineered for silent operation or employing topologies such as quasi-resonant converters, which are soft-switching DC-DC converters that activate and deactivate their switches (usually MOSFETs) under zero voltage or zero current conditions. This converter substantially diminishes electromagnetic interference, switching noise, and component stress (Ahamed et al., 2024).

b) Employment of noise filters

Ferrite chokes, cores, and beads can be utilized to mitigate electromagnetic interference (EMI) by transforming electrical disturbances into heat, so substantially diminishing or eradicating noise from the system.

c)Consideration of design of the system

When developing the solar photovoltaic cooker, integrate requisite filters and provide adequate grounding to mitigate noise transmission over the power or control lines.

4. DESIGN OF SOLAR PV E-COOKING SYSTEM WITH THE AID OF CONVERTERS

i)Induction Heating

Induction heating e-cooking system(Figure.5) utilizes an induction coil, powered primarily by solar panels to ensure energy sustainability. The Half-Bridge inverter topology enhances efficiency by allowing for precise control of the power delivered to the induction coil. A battery charging and power management device is integrated into the system to regulate output power and maintain optimal performance, adapting to various cooking settings by selecting appropriate operating frequencies. This configuration not only maximizes energy usage from solar resources but also provides a backup through mains electricity, ensuring reliable performance for cooking applications regardless of solar availability.

Figure 5. Induction heating e-cooking system

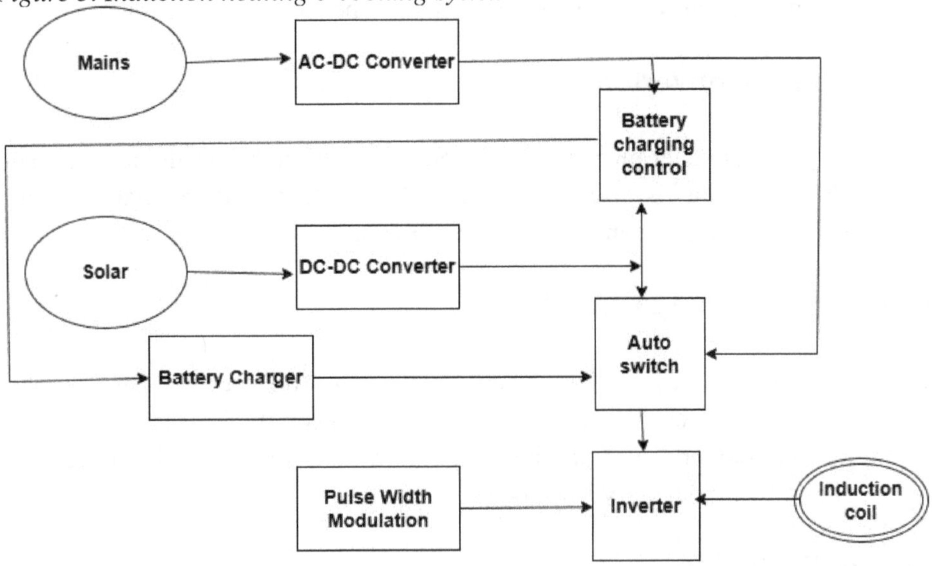

ii.AC to DC Converter

The AC to DC converter efficiently transforms the mains electricity into a stable DC output, which is essential for both battery charging and the operation of the Half-Bridge converter in the induction heating system. By providing a consistent DC supply, the boost converter can effectively increase the voltage to the required level for battery charging, ensuring that the energy storage is both reliable and efficient. This arrangement allows the system to seamlessly integrate the mains power with solar energy, optimizing the overall performance and enabling the induction coil to operate effectively under varying cooking settings. The result is a flexible and efficient cooking solution that leverages available energy sources while maintaining compatibility between component

iii. Solar

The solar energy serves as principal source of power for culinary. Solar technology consists of solar panels that turn solar energy into direct current. Photovoltaic modules transform solar energy into electricity (Nandha Gopal et al., 2024). Output power of a photovoltaic system is not consistently steady during the day due to its non-linear features. This kind of system integrates battery and a DC-to-DC converter with photovoltaic systems to manage their voltage. This research identifies solar energy as the primary source of energy for heating the induction coil. This solar panel system relies on solar radiation to generate energy.

iv. DC to DC Converter

The voltage converter manages the uncontrolled DC output from the solar panel. A stable DC supply is essential to charge the unit to provide regulated DC for charging and to deliver a consistent power to the inverter.

v. Battery Charging Circuit

Batteries store energy and have electrical cells that work together. Battery chargers that are small enough to carry around serve to keep the load level (Mitali et al., 2021). The charger driver circuit manages how much the battery is charged and discharged so that the battery lasts longer.

vi.Auto Switching

By analyzing power usage, the auto-select switch automatically chooses the best power source from among the solar panels and the power grid. Solar power is the most cost-effective option that was looked at. Microcontroller software controls the auto switch. The amount of power in the battery is compared to the chosen switching state. If the battery level is high enough, the battery powers the coil. This auto switch is set up so that the source is chosen at the start of each cooking state. The mains power will be used if the battery isn't enough to provide the power needed for the culinary state chosen and the time period estimated. Either the solar panels or the electrical system could be used to charge the battery.

5. REAL WORLD EXAMPLES SOLAR PV E-COOKING SYSTEM

5.1. Case Study 1: Smart Clean Cooking system

The majority of existing sustainable cooking options entail using biogas, ethane, LPG, and biomass-based fuels, all of which release carbon dioxide and harmful pollutants. On other hand, the main drawbacks of solar power cooking appliances include their expensiveness and intricate technology, which restrict their adoption with limited resources. Batteries are necessary for both off-grid and conventional solar powered homes to operate. Majority of solar goods are modest and only capable of recharging phones and lighting lights because battery packs have the lowest lifespan of any component, pose security hazards, are composed of hazardous materials, and can constitute up to 70% of the system's overall cost. Solar electric clean cooking systems are being produced by corporations; however their expensive price tag makes them unaffordable for those with low incomes. By generating electricity off the grid to operate electrical pots and pans or induction cook stoves without the need for batteries, the Smart Clean Cooking System lowers costs and complexities in technology, giving consumers a economical initial move toward energy independence as shown in Figure.6. By meeting their present and prospective electrical requirements in accordance with individual and environmental restrictions, battery-optional solutions create an innovative, potent first-rung energy system that enables consumers to realize their goals. Up to 750W of electricity can be delivered by individual modules, and the system can function in two ways: 1) Direct-Drive: This method uses solar photovoltaic panels to supply loads with usable DC electricity. 2) Solar Battery Charge Controller: By including battery storage into the same system, electricity may be accessed even when sunlight isn't available. This device also has the ability to switch back to Direct-Drive as a backup plan in the event that the battery fails. The costliest, troublesome, and short-lived part needed by conventional off-grid energy systems is now an add-on thanks to this technology.

Figure 6. Smart clean cooking system

Plug-and-play solutions have several advantages for both the environment and end users, and they aim to achieve several sustainable goals: 1) Energy Conservation and Efficiency: This approach promotes energy-saving behaviours, such as making effective use of solar energy when it is available. End users may install a smaller battery to increase the amount of electricity available if solar energy is harnessed without batteries. 2) Electronic Waste Reduction: Our battery-optional, scalable solution extends system working and service life by growing with end users' energy demands. Very few proprietary parts are used in these systems, and the circuitry is made to be disassembled for recycling, reusing, and repair. 3) Economic Development: This technology up skills and employs men and women across the supply chain, including distribution, sales, installation, repair, and reclamation, while empowering micro, small, and medium-sized businesses and small-scale farms for local economic growth. 4) Conserve Time, Money, and Trees: Gathering solid fuels like wood can take up to five hours of a day, typically for women; traditional, filthy cooking fuels like charcoal can cost up to 76% of daily revenues; and deforestation is accumulating on the continent, affecting hectares per year. 5) Health and Gender Equity: Clean cooking improves women's quality of life and expands opportunities for female entrepreneurs while lowering health problems brought on by indoor air pollution.

5.2. Real World Examples With Case Study 2

5.2.1. Utilization of Artificial Intelligence in Solar Photovoltaic Systems

Recent decades have seen the majority of research advancements in solar power conservation. The stability of the entire system is problematic because of the unknown nature of solar panel components. More appropriate optimization techniques than traditional ones have been proposed in earlier research. In solar energy usage, the optimization algorithms have demonstrated excellent performance with regard to of dimension, peak demand, and generation of electricity. Additionally, the optimizations help to improve peak power incorporation, reduce loss of power and maintenance costs, and determine the optimum way to govern the process.

5.2.2. Surveillance of Photovoltaic Systems

The monitoring system for PV plants keeps track of several electrical generating indications and faults in addition to gathering and analyzing a number of parameters related to the reliability and stability of any PV facility. Over the past ten years, research has described many PV monitoring system elements. This includes a comprehensive analysis of the main PV Monitoring assessment techniques and their respective effectiveness. A PV monitoring system's key components include sensors and their use manuals, as well as controllers used for data gathering, delivery, storage, and evaluation. To build practical, economical, and successful PV monitoring systems for medium- to small-sized solar panels without sacrificing the necessary efficiency, all of these variables must be considered. PV monitoring systems are made to guarantee that data on energy production, operational temperature assessment and loss of energy related to different potential problems are available. Consequently, a great deal of work has been done to create effective PV monitoring systems. A variety of industrial goods that have been integrated into the surveillance system could be included. Given the growing number of commercial products based on multiple ideas, it is essential to research how each one works and what its features are. Choosing appropriate solution for a given climate is essential for a successful PV tracking. In this sense, space-based technologies might be more affordable due to the absence of sensors. The unfavourable aspect of space-based systems is their low precision that is greatly impacted by meteorological conditions. Therefore, only ground-reliant systems are included in this analysis of PV monitoring systems.

Failure to promptly identify and rectify flaws in components of solar photovoltaic systems (hybrid PV, stand-alone, or grid-connected) could bang the efficiency and power generation of the system. Furthermore, certain faults might exacerbate the risk of fire outbreaks. Fault detection and diagnosis systems are essential for the rigidity, execution, and security of solar photovoltaic systems. When a photo voltaic module malfunctions, the issue is generally associated with the product's protection. Photovoltaic system defects lead to unforeseen safety issues, diminished performance, power availability challenges, and compromised integrity of the system. Numerous studies have recognized and investigated the types of PV failures. Issues include fading, bursting, deterioration of the anti-reflective covering, swelling, and markings, rusting of the bus bars, break encapsulations at cells and joints, and loss of backside adhesion. . PV module problems are often classified into categories: permanent and transitory. Permanent defects encompass cell turning yellow, abrasions, delamination, and burned cells. Consequently, these problems can be averted by merely substituting the impaired modules.

5.2.3. Solar Photovoltaic Generator Modelling

Grid-connected solar renewable energy installations have become prevalent in India. The majority of these exhibit electrical outputs varying from hundred kilowatts to megawatts. Small grid-connected solar power plants aren't being built in India, as is the case in numerous other nations. The capacity of solar PV power plants to produce energy has surged significantly in recent years, Solar PV power generation, historically the predominant operational mode, was considered the exclusive option for solar power plants until a few years ago. The power plant functioned as an electric power generator for local demands in this instance. Due to the limitations of solar power generation based on sunlight availability, battery-based energy storage systems are frequently employed.

6.POTENTIAL STUDY PROSPECTS FOR THE E-COOKING INDUSTRY

Electric culinary inventions like electric pots and induction stove, exhibit significant potential due to their exceptional energy efficiency. This aims to enhance system efficiency by integrating these technologies along with green sources of solar power. This explains why this is the situation. The research indicates the amalgamation of electric cooking inventions including solar photovoltaic systems and microgrids might be price-controlled, hence enhancing way in to green cooking for a larger population in rural locations. Obstacles remain, including the elevated initial cost, system dependability, and restricted capacity during peak demand periods. Nonetheless, analogous challenges are being faced now. Proposed methods include demand-based techniques, innovative pricing structures, intelligent battery management, and optimized hybrid mini-grids. The issue of cost is a significant hurdle for consumers regarding the adoption of e-cooking powered by renewable energy technology. Various socio-cultural problems, such as preferences, knowledge levels, gender dynamics, and impacts on livelihoods, contribute to this difficulty. The source of electricity for the electronic cooking system, whether renewable or not, is irrelevant; the former significantly reduces greenhouse gas interior pollution of air and toxins in comparison to conventional propellant like kerosene ans LPG. This can be true irrespective of whether the electronic cooking equipment is fueled by renewable energy.

Consequently, it is vital to possess a thorough comprehension of local surroundings, to adopt the technique of "fuel stacking," and to execute tactics for behavioral modification. Additional research is required to examine the technical, economic, social, and environmental dimensions of these technological solutions to expedite the adoption of e-cooking solutions, aid in achieving sustainable energy access objectives, and support the realization of clean cooking targets.

6.1. Progress in the Design of E-Cookers

Further prospect solutions in this domain encompass the amalgamation of automation field and the Internet of Things (IoT), alongside the enhancement of solar e-cookers with the artificial intelligence (AI). These represent only a subset of the potential solutions that exist. To improve the efficiency of solar electric cookers utilizing optimization approaches, artificial intelligence methods such as deep learning and machine learning may be employed. By employing these approaches, adaptive systems may be constructed. To achieve optimal performance, efficiency, and usability, these systems would continually monitor and assess many factors, including as sun irradiation, ecological temperature, customer behavior, for varying settings of cooker accordingly. This would enable the cooker to attain optimal performance, efficiency, and usefulness. It is viable to up skill machine learning model to predict culinary performance under various settings by enabling them to learn from historical data. The implementation of such a system would enable real-time alterations to components, including the angle of the reflectors and the placement of the cooking pot. Deep learning technique may be adopted to analyze pictures or data of sensors to identify probable issuances, like misalignments or obstructions, and to yield solutions to these challenges. This would be utilized to furnish further information.

Additionally, reinforcement learning techniques may be explored to develop self-optimizing systems capable of learning from events and modifying their behavior to enhance performance metrics. This may be achieved through the integration of reinforcement learning systems. Factors like as cooking duration, energy efficiency, and user convenience may all be addressed in this context. Internet of Things (IoT) and automation technologies may be integrated into electronic cookers to improve the cooking experience, promote user convenience, and increase the appliance's energy efficiency. Culinary tasks may be remotely watched, supervised, and automated due to recent technological advancements. This would result in a decrease in the necessity for human involvement, hence facilitating a more effective utilization of existing resources. It is possible to equip Internet of Things-enabled electronic cookers with sensors that monitor many factors, including temperature, energy consumption, and the progression of the cooking process accordingly. This is an achievable task. The data may be sent to a central control system or a user-operated mobile device to facilitate real-time monitoring and remote management of the cooking process. This is conducted to facilitate remote control and oversight of the cooking process. User end may get acknowledgements, configure settings, and initiate or terminate the cooking process remotely using a remote control. This enhances the device's adaptability and convenience, facilitating its usage. Certain portions of the cooking process may be automated using technologies such as programmable logic controllers and microcontrollers, which are well-suited for electronic cookers. For example, cooking profiles dependent on temperature and duration may be acquired over time or pre-programmed, ensuring consistent and flawless outcomes while also minimizing wasteful energy consumption.

6.2. Grid Integration and Energy Management

This part will talk about two possible areas for future research: how to connect electric cookers to smart grids and how to study demand-side control systems. Adding electric cookers could make smart grid systems better by increasing energy efficiency, managing demand, and keeping the grid stable, which could lead to benefits (Arun sampul Thomas et al., 2023). Smart grid technologies, such as demand-side management methods and better tracking infrastructure, let electric devices talk to the grid in both directions, which makes demand-side management possible. Electric stoves can also have set up to briefly lower their energy use in reaction to system emergencies. This can be done through actual price signals or by moving its maximum load to off line-peak periods. Additionally, e-cookers can briefly lower the amount of energy they use. These projects that increase demand-side flexibility could assist remaining supply with demand, make the grid more stable, and make it easier to use sporadic green energy sources. Smart grid technologies could also make it easier to use agile price models, which would encourage people to change when they cook to times when power demand is lower or when green energy is more available. The investigation also looks into how demand-side control tools, such as load shedding and time-of-use pricing, affect the use of electric cooking and the security of the grid. These methods can change how customers act and how much energy they use, which can have an effect on how profitable and long-lasting e-cooking options are overall. Time-of-use pricing, which means charging more during times of high demand, might encourage people to move their food operations to times when demand is lower, which would make the grid less stressed. On the other hand, how well this approach works may depend on a number of factors, such as how well customers understand it, how much they think it costs, and how easy it is to find other cooking choices during busy times. Load shedding, a method by which energy supply to certain areas or users is briefly reduced or stopped during high demand times, may also have an effect on the use of electronic cooking. Some load shedding may be necessary to keep the power grid stable, but customers may not use e-cooking options because they are worried about how reliable and easy they are to use during frequent or long-lasting load shedding. In order to figure out how these methods work, in-depth studies that include polling consumers, analysing energy use data, and grid models may be carried out. Also, field trials and test projects have a big chance of teaching us a lot about how people act and how demand-side control measures really work when it comes to electric cooking and keeping the grid stable.

6.3. Socio-Economic and Gender Considerations

Some potential areas for further investigation in this category include investigating the impact of gender on the adoption of e-cooking and determining whether the concept can be economically viable. In rural areas and low-income communities, researchers can evaluate the financial viability of e-cooking options over the long term. This evaluation may provide additional consideration to the tariff regulations, support programs, and product pricing for the groups that are being examined more closely. The cost of e-cooking options is significantly influenced by tariff policies. In locations with progressive pricing systems, where fees increase in accordance with utilization, individuals with low incomes may encounter challenges when employing electronic cookery. Conversely, the cost of electric cooking may be reduced and its popularity may increase as a result of reduced prices for electric cooking apparatus or the abundance of significant benefits associated with electricity. The potential environmental advantages, as well as the expenses associated with gas and maintenance must be examined in these investigations. At present, the primary objective is to establish laws and programs that are inclusive of all individuals. However, experts can also gain a deeper understanding of the ways in which gender influences energy access and the ways in which socioeconomic factors influence the degree to which individuals embrace e-cooking. Women typically assume responsibility for household tasks such as energy management and preparation. As a result, their perspectives and experiences are highly beneficial in the development of solutions for e-cooking. At this kind of scenario, Qualitative evaluation techniques like meeting, interviews, and engaging methods could be applied to ascertain the concerns, preferences, and anxieties that men and women have regarding the acceptability of e-cooking. These findings have the potential to facilitate the development of computer culinary tools, programs that enhance individuals' abilities to address the needs and issues of women, and awareness campaigns. The acceptability of e-cooking options may also be significantly influenced by socioeconomic factors, including income, education, cultural norms, familial relationships, and income range.

6.4. Behavioural Change and Adoption Strategies

This field develops economic models and theoretical frameworks to support electronic cooking. These therapies can be directed and their underlying concepts elucidated by theoretical frameworks and models. Theories and models from behavioural economics, sociology, and psychology can elucidate and affect the acceptability of e-cooking. Community-based participatory initiatives can alter behavior as the impacted individuals plan and implement them. Information may be disseminated and e-cooking concepts promoted via established social networks, community leaders, and cultural norms. Future research may explore pay-as-you-go or mobile-enabled fee-for-service approaches to enhance the affordability and accessibility of electronic cooking gadgets and services.

Pay-as-you-go solutions remove the initial cost obstacle by allowing users to finance electronic cooking equipment and services through manageable monthly payments. These methods, integrated with utility billing systems or supported by mobile money networks, can offer electric cooking assistances through a money paid mechanism, avoiding substantial initial expenditures. Another research focus may be fuel stacking systems, wherein households integrate conventional and electric cooking fuels for various meals to optimize e-cooking at reduced expenses. Fuel stacking recognizes that numerous households are hindered from transitioning to e-cooking due to cultural, infrastructural, and budgetary limitations. Integrating traditional and contemporary fuels enables households to optimize both systems while mitigating their adverse impacts. Families may prepare expedited meals or temperature-controlled delicacies with electronic kitchen gadgets. Residences may still require traditional energy for boiling water and preparing large quantities of meals. In comparison to conventional cooking methods, e-cooking enhances indoor air quality, reduces labour requirements, increases convenience, and conserves energy and costs. Assessing fuel stacking solutions necessitates an understanding of family energy use, culinary methods, and individual choices. Energy management and auditing, and ethnology observation can be employed to assess fuel stacking across various contexts and ascertain its potential advantages.

CONCLUSION

Although culinary technologies have grown more quick, easy, and effective, it is still very difficult for global governments and lawmakers to guarantee that they are available, healthy, dependable, and reasonably priced. Right now, the push for renewable power represents highest point and requires immediate action. Since photovoltaic culinary suits perfectly in terms of simplicity, oversight, cost-effectiveness, cleanliness, and compactness, it could represent practical way to meet culinary requirements. Because of the realization that solar PV cooking solutions offer a good way to meet cooking requirements in cities and rural settings. It may expand considerably with effective promotion, and it won't be long before it turns into among the popular cooking methods. It may expand considerably with effective promotion, and it won't be long before it turns into popular techniques for cooking. Additionally, the investigation methodically looked recent advances and patterns in solar PV-based e-cooking technologies, showing how they could help green cooking issues, especially in areas that lack electricity availability. The results demonstrate interdisciplinary nature of this field, which includes photovoltaic -hydrogen frameworks, PV-integrated appliances for cooking, and uptake techniques that are accessible across societal groups. The next research ought to canter on creating novel heat and blended storage solutions, incorporating AI and I0T for forecasting structure optimization, exploring scalable hydrogen-based cooking technologies, performing social and economic analysis of customer behavior, and evaluating the financial and environmental effects through life-cycle assessments. The revolutionary significance that photovoltaic e-cooking plays in advancing ecological cooking and environmental sustainability is highlighted in this assessment.

REFERENCES

Ahamed, S., Awalin, L. J., Hasan, S. M. N., Saha, A., Ali, M. S., Syafiq, A., & Wang, L. (2024). Quasi resonant topology based highly efficient solar-powered induction cooker. Bulletin of Electrical Engineering and Informatics, 13(6), 4506-4521.

Ali, A. O., Elgohr, A. T., El-Mahdy, M. H., Zohir, H. M., Emam, A. Z., Mostafa, M. G., & Elhadidy, M. S. (2025). Advancements in photovoltaic technology: A comprehensive review of recent advances and future prospects. *Energy Conversion and Management*, *X*, 100952.

Ansari, S., Ayob, A., Lipu, M. S. H., Saad, M. H. M., & Hussain, A. (2021). A review of monitoring technologies for solar PV systems using data processing modules and transmission protocols: Progress, challenges and prospects. *Sustainability*, *13*(15), 8120.

Arun Sampaul Thomas, G., Muthukaruppasamy, S., Saravanan, K., & Muleta, N. (2023). Revolutionizing Smart Grids with Big Data Analytics: A Case Study on Integrating Renewable Energy and Predicting Faults. In *Data Analytics for Smart Grids Applications—A Key to Smart City Development* (pp. 179–198). Springer Nature Switzerland.

Batchelor, S., Talukder, M. A. R., Uddin, M. R., Mondal, S. K., Islam, S., Redoy, R. K., & Khan, M. R. (2018). Solar e-cooking: A proposition for solar home system integrated clean cooking. *Energies*, *11*(11), 2933.

Damayanti, R. W., Setiadi, H., Laksono, P. W., Rizky, D. L., & Entifar, N. A. (2023). Factors affecting technological readiness and acceptance of induction stoves: A pilot project. *Emerging Science Journal*, *7*(6), 1892–1923.

Delapedra Silva, V., Ferreira, P., Cunha, J., & Kimura, H. (2022). Methods for financial assessment of renewable energy projects: A review. *Processes (Basel, Switzerland)*, *10*(2), 184.

Dessart, F. J., Barreiro-Hurlé, J., & Van Bavel, R. (2019). Behavioural factors affecting the adoption of sustainable farming practices: A policy-oriented review. *European Review of Agriculture Economics*, *46*(3), 417–471.

Felix, P. G., Rajagopal, V., & Kumaresan, K. (2025). Characterisation of erythritol as a potential phase change material. *Australian Journal of Mechanical Engineering*, *23*(2), 205–219.

Joshi, S. B., & Jani, A. R. (2015). Design, development and testing of a small scale hybrid solar cooker. *Solar Energy*, *122*, 148–155.

Leach, M., Mullen, C., Lee, J., Soltowski, B., Wade, N., Galloway, S., & Batchelor, S. (2021). Modelling the costs and benefits of modern energy cooking services—Methods and case studies. *Energies, 14*(12), 3371.

Leary, J., Leach, M., Batchelor, S., Scott, N., & Brown, E. (2021). Battery-supported eCooking: A transformative opportunity for 2.6 billion people who still cook with biomass. *Energy Policy, 159*, 112619.

Łosiewicz, B. (2024). Recent Advances in Technologies of Green Hydrogen Production. *Energies, 17*, 4514.

Misra, N., Anand, A., Pandey, S., Kant, K., Shukla, A., & Sharma, A. (2023). Box-type solar cookers: An overview of technological advancement, energy, environmental, and economic benefits. *Energies, 16*(4), 1697.

Mitali, J., Dhinakaran, S., & Mohamad, A. A. (2021). Energy storage systems: A review. *Energy Storage and Saving, 1*(3), 166–216.

Mukelabai, M. D., Wijayantha, K. G. U., & Blanchard, R. E. (2022). Hydrogen for cooking: A review of cooking technologies, renewable hydrogen systems and techno-economics. *Sustainability, 14*(24), 16964.

Nandha Gopal, J., Muthukaruppasamy, S., Rajagopal, V., Thomas, G. A. S., & Parimalarsundar, E. (2024). Artificial Intelligence Governance and Comprehensibility in Renewable Energy Systems. In Explainable AI (XAI) for Sustainable Development (pp. 73-88). Chapman and Hall/CRC.

Panagoda, L. P. S. S., Sandeepa, R. A. H. T., Perera, W. A. V. T., Sandunika, D. M. I., Siriwardhana, S. M. G. T., Alwis, M. K. S. D., & Dilka, S. H. S. (2023). Advancements in photovoltaic (Pv) technology for solar energy generation. Journal of research technology & engineering, 4(30), 30-72.

Paneru, B., Paneru, B., Alexander, V., Nova, S., Bhattarai, N., Poudyal, R., . . . Boland, J. J. (2024). Solar energy for operating solar cookers as a clean cooking technology in South Asia: A review. Solar Energy, 283, 113004. Chiloane, L. L., Aswat, M., Yen, Y. C. J., & Cronje, W. A. (2023). A PV-supplied cooking solution using a hybrid electric-thermal energy storage system. In 2023 31st Southern African Universities Power Engineering Conference (SAUPEC) (pp. 1-5). IEEE.

Prasanna, U. R., & Umanand, L. (2011). Optimization and design of energy transport system for solar cooking application. *Applied Energy, 88*(1), 242–251.

Rashid, F. L., Al-Obaidi, M. A., Mahdi, A. J., & Ameen, A. (2024). Advancements in Fresnel lens technology across diverse solar energy applications: A comprehensive review. *Energies, 17*(3), 569.

Redekar, A., & Deb, D. (2024). Design of an advanced solar cooker with integrated self-powered active sun tracking and dual heating facilitated for rural communities. [IJIDeM]. *International Journal on Interactive Design and Manufacturing*, 1–25.

Ronge, H., Niture, V., & Ghodake, M. D. (2016). A review paper on utilization of solar energy for cooking. *Imperial Int J Eco-friendly Technolo*, *1*, 121–124.

Sadeq, A. M., Homod, R. Z., Hussein, A. K., Togun, H., Mahmoodi, A., Isleem, H. F., & Moghaddam, A. H. (2024). Hydrogen energy systems: Technologies, trends, and future prospects. *The Science of the Total Environment*, 173622. PMID: 38821273

Sarangi, A., Sahoo, S., Nayak, J., & Mallik, R. K. (2024). Advancements and global perspectives in solar cooking technology: A comprehensive review. *Energy Nexus*, *13*, 100266.

Saxena, A., Norton, B., Goel, V., & Singh, D. B. (2022). Solar cooking innovations, their appropriateness, and viability. *Environmental Science and Pollution Research International*, *29*(39), 58537–58560. PMID: 35778663

Scott, N., Leach, M., & Clements, W. (2024). Energy-Efficient Electric Cooking and Sustainable Energy Transitions. *Energies*, *17*(13), 3318.

Teske, S., Niklas, S., & Miyake, S. (2023). Technical Scenario for 100% Renewable Energy in Nepal by 2050: Possible Transition Pathways for NDC & LTS Implementation; prepared for World Future Council, WWF and Brot für die Welt by the University of Technology Sydney, Institute for Sustainable Futures; February 2023.

Chapter 7 Synthesis of Solar Cookers and Evaluation of Their Thermal Performances.

Abdennabi Brahmi
https://orcid.org/0009-0004-4990-645X
Faculty of Sciences and Techniques, Moulay Ismail University, Errachidia, Morocco

Iftissane Elmustapha
Faculty of Sciences and Techniques, Moulay Ismail University, Errachidia, Morocco

Jaouad Foshi
Faculty of Sciences and Techniques, Moulay Ismail University, Errachidia, Morocco

Ahmed El-Abbassi
Faculty of Sciences and Techniques, Moulay Ismail University, Errachidia, Morocco

Mohammed Ali Arbaoui
https://orcid.org/0009-0002-5945-1036
Faculty of Sciences, University Mohammed V, Rabat, Morocco

ABSTRACT

The main objective of this work is to provide an overview of used solar cookers and to describe the essential parameters for testing and evaluating their thermal

DOI: 10.4018/979-8-3373-2535-4.ch007

performance. Several categories of solar cookers are found in the literature servey. The price of solar cookers has recently been reduced by around 56%, from 530 Dhs to 235 Dhs. The manufacturing cost remains around 200 Dhs. They are classified into four main families: box cookers, concentrator cookers, ovens, and indirect solar cookers. Each family is divided into three types of solar cookers namely: Photovoltaic, Thermal and Hybrid solar cookers. Photovoltaic solar cooker remains the cooker of the future. Indeed, the cooking temperature can be increased by adjusting the parameters of the PV generator used. Similarly, the electrical energy provided by the PV source can be stored using solar batteries. This storage allows the users to prepare meals regardless of lighting conditions. In addition, the cost of PV modules is decreasing.

I. INTRODUCTION

The exhaustion of fossil fuel resources is inevitable given the high global consumption of energy in its various forms (Panwar N. et al., 2012). This has led to the search for new energy sources known as renewable energies (Khennoufa A. and Lachibi O., 2017). These new sources include wind, biomass, tides, sea and ocean waves, heat from the Earth's depths, and the sun (Panwar N. et al., 2012).

The latter source deserves to be at the top of this list of sources because of its abundance, but also its distribution in nature more than any other renewable source (Yettou F. et al., 2012).

Morocco boasts a significant amount of sunshine. This represents significant potential and a practical solution to the growing challenges in the energy and environmental sectors (Bouhamida M. and Henichi I., 2020).

In fact, it is a primary, urgent, and shared objective for all humanity to preserve a sacred environment for future generations (Yakhlef Y. and Mehdaoui A., 2018). This encourages and requires the development of all kinds of solar applications.

Solar cooking is one of the most attractive applications (Panwar N. et al., 2012). It's an alternative for conserving our forests, limiting the misuse of fossil fuels, which emit greenhouse gases that threaten to jeopardize the future of humanity, and also reducing cooking costs. Solar cookers are well-known worldwide. Cooking is a regular and frequent human activity. It allows for the preparation of healthy foods. In areas where butane gas supplies are difficult to access, residents often use wood to cook their food (Djebara A. and Guettaf Z., 2018).

Solar cooking is emerging as a key alternative for meeting vital needs without affecting plant biodiversity in forests while reducing greenhouse gas emissions (Mullick S. et al., 1996).

Solar cookers are a technology that provides heat in a natural and renewable way. This technology often uses two techniques: concentrating the sun's rays and harnessing the greenhouse effect (Panwar N. et al., 2012).

Concentrating the sun's rays onto a collector surface reduces the effective receiving area and thus minimizes heat loss at the receiver. As a result, the temperature reached after conversion to heat is higher than with devices using non-concentrating collectors. This type of cooker uses reflective surfaces based on mirrors (Mbodji N. and Hajji A., 2017).

In this context, the work presented in this chapter addresses the history of solar cookers, their overview, their specific features, and the relevant characterization and evaluation parameters for this technology. This chapter is organized according to the following outline with three main sections:

The first section presentes a brief history of the evolution of this technology and its literature survey.

The second section describes the principle of working and basic models of solar cookers, conditions of use and a study of general information about solar cookers.

The third section discusses the definitions, and formulas for the various important parameters used for the practical evaluation of the thermal performance of these solar cookers. In the last, the most advantages of this techlogy are cited.

Finally, a conclusion containing the interpretations and further development considered will conclude this work.

II. GENERALITY OF SOLAR COOKERS

1. State of the Art of Solar Oven

Solar cooking is far from modern (Funk P., 2000). In 1945, Sri M. Kghosh designed the first box-shaped solar cooker as a commercial product. In 1950, Indian scientists designed and manufactured several solar cookers with solar reflectors (Patel N. and Philip S., 2000).

Swiss physicist Horace Bénédict built a solar cooker in 1767. This researcher constructed miniature greenhouse of five nested glass boxes placed on a black tray (Nahar N., 2003). The deepest box cooked well. The temperature rose to 109°C.

In 1976, two American researchers, Brbara Kerr and Sherry Cole, built a series of box-type solar cookers that were easy to implement at home (Mosbah and A. Achour A., 2020).

Richard Petela designed and developed a parabolic trough solar cooker in 2002 as shown in figure 1-a below. He discussed his theoretical analysis and presented the energy loss analysis. He also analyzed the exegetics of the surface exposed to light rays, which absorbs the radiant flux at different temperatures (Mosbah and A. Achour A., 2020).

During the same year, researcher Oztuk experimentally determined the energy and exegetical efficiencies of this type of cooker by applying international standards (See figure 1-b in the next page).

In 2003, Philip and Sonune developed a Fresnel-type concentrating solar cooker for domestic use as representing in figure 2 -a. This type provides the right temperature for cooking and frying (Kassmi K., 2021).

A year later, Franco et al. improved this type of cooker with a few modifications. They added an insulated box to hold the container as shown in figure 2 –b (See next page). The addition of this box further improves cooking capacity by using an optimized absorber.

In 2007, José M. Arenas developed a new portable solar cooker made from a large foldable parabolic reflector as shown in figure 3 –a (Next page).

Tests and trials showed that this type of cooker could reach an average power of 175 W with an efficiency of approximately 26.6% (Ashok K . and Sudhir C.V., 2009).

It could cook a simple meal for two people in two hours.

Figure 1. Parabolic trough solar cooker: a) Cylindrical-parabolic cooker studied by Richard Petela. b) Representation of the parabolic trough cooker tested by Ozturk.

a

b

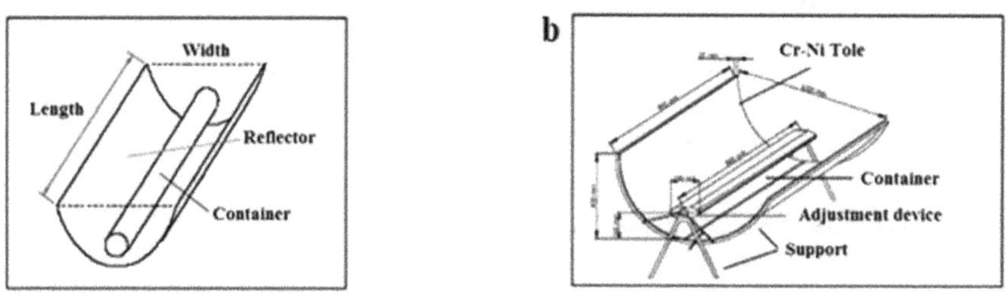

Figure 2. Fresnel-type concentrating solar cooker: a) Parabolic cooker with Fresnel mirrors made by Sonune and Philip. b) Parabolic cooker composed and produced by Franco et al.

a

b

In 2009, Crupp et al. developped a parabolic solar cooker (Figure 3 -b). It contained a device that recorded ambient temperature, cooking temperature, and solar irradiation (Panwar N. et al.2012).

In 2004, Esen proposed and demonstrated the feasibility of an indirect solar cooker containing refrigerants (Figure 4 -a). This cooker allows for cooking and/or storing hot items (Ashok K . and Sudhir C.V., 2009). It is based on an evacuated tube solar collector with a long integrated heat pipe leading directly to the hob (Yettou F. et al. 2012).

In 2005, Sharma et al. evaluated the thermal performance of such a solar cooker based on an evacuated tube collector and a latent heat storage unit equipped with phase change materials (PCM). This configuration is illustrated in figure 4 -b in the next page.

In 2004, Esen proposed and demonstrated the feasibility of an indirect solar cooker containing refrigerants (Figure 4 -a). This cooker allows for cooking and/or storing hot items (Ashok K . and Sudhir C.V., 2009). It is based on an evacuated tube solar collector with a long integrated heat pipe leading directly to the hob (Yettou F. et al. 2012).

Figure 3. New portables solar cookers: a) Umbrella portable solar cooker. b) Portable parabolic solar cooker.

a

b

In 2005, Sharma et al. evaluated the thermal performance of such a solar cooker based on an evacuated tube collector and a latent heat storage unit equipped with phase change materials (PCM). This configuration is illustrated in figure 4 -b in the next page.

Figure 4. Indirect solar cooker: a) Schematic of cooker containing refrigerants. b) Solar cooker with phase change materialas (PCM).

a

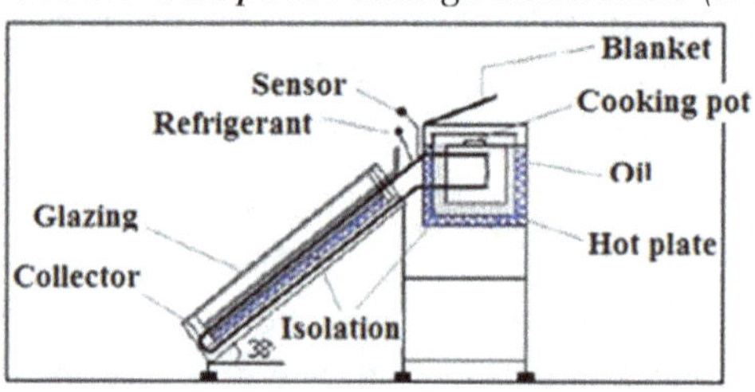

b

In 2010, Al Soud et al. developped and implemented a parabolic cylindrical solar cooker based on a solar tracking system as shown in figure 5 below. The latter is based on a microcontroller to manage the cooker's movement.

Figure 5. Cylindro-Parabolic Cooker with Tracking Mechanism.

In 2011, Abu-Malouh et al. built a parabolic solar cooker with a dual-axis tracker (See figure 6). This cooker concentrates solar radiation onto a container attached to the top of the cooker while tracking the sun's path.

Figure 6. Parabolic solar cooker with tracking system.

2. Projects of Solar Cookers

In many countries, wooded areas are degraded and access to heating or cooking energy is difficult and expensive. It is with this in mind that some governments and non-governmental organizations are promoting the use of solar cookers and ovens (Shukla S. K., 2009).

Among these actors, more than 300 non-governmental organizations NGO, businesses, educational institutions, manufacturers and governments have come together to form the Solar Cookers Network. Its aim is to improve health, the economy and the environment through collective actions around solar cooking, water pasteurization and food processing, particularly in the most needy regions.

One of the most active French non-governmental organizations in the world is the Bolivia Inti – Sud Soleil association. Since its creation in 1999, it has enabled the installation of nearly 10,000 ecological cookers in the Andes, used every sunny day by 50,000 people. This association relies on the commitment of more than 2,500 members. Open to other territories, the Bolivia Inti – Sud Soleil association has initiated, in partnership with other associations, programs to evaluate and disseminate ecological cookers in Africa.

Other non-governmental organizations are involved in the promotion and dissemination of solar cookers, here is a list of French non-governmental organizations:

- Laboratory of Energy and Materials,
- Sun benefits,
- Solar Cooking Concept,
- Other non-governmental organizations are involved in the promotion and dissemination of solar ovens, here is a list of French non-governmental organizations,
- Agrometeorological Applications Associates,
- Synopsis.

3. Essential Factors for Cooking

The solar cooker should be placed outdoors, in a sunny, windy location, but protected from strong winds, clouds, mist, dust, and possible shade from nearby trees or buildings, which reduces radiation and increases cooking time (Shukla S. K., 2009).

The chosen location must be far from human waste, animal waste, and sewage disposal sites.

The solar cooker should be used when the length of the shadow on the ground is shorter than the actual height of the cook.

- Cooking time varies and will depend on the amount of sunlight.
- Cooking times are much longer than traditional cooking methods; you need to plan ahead and not be in a hurry.
- Cooking outdoors depends on the amount of sunlight; it's not possible every day or at all hours.
- It's not suitable for frying or grilling, which require high temperatures.

4. Advantages of Solar Oven Technologies

Cooking with a solar cooker has many advantages (Funk P., 2000, Mosbah and A. Achour A., 2020) which are:

- It does not contribute to greenhouse gas emissions. It does not irritate eyes because it does not produce smoke.
- This cooker technology uses inexhaustible, clean, and free energy.
- This cooking technique preserves meals better than cooking at high temperatures. Cooking at low temperatures helps preserve some of the vitamins and minerals in food.
- Cooked food is very tasty. This cooking method, which does not require the addition of fat, is more dietetic than cooking with added fat.
- Solar cookers are practical, made from lightweight materials, and can be transported anywhere.
- Solar cookers offer the possibility of cooking outdoors for a large part of the year.
- When cooking in the solar cooker is in progress, you have your time for other activities because your presence is not required.
- Solar cooking allows water to be pasteurized.
- Pots are easy to maintain because food cooked at low temperatures does not stick. The oven is very easy to maintain.
- Commercially available solar cookers have a long life span.
- Solar ookers can be made for a small fee.

Although these solar cookers have many interesting advantages, this technology is not without its drawbacks (Panwar N. L. et al., 2010). For examples, some of the downsides of these solar ovens are:

- Solar cookers take long time to cook food compared to an oven.
- Solar cookers are less usable in cloudy or rainy weather, so some fuel-based backup heat source must still be available to cook food at these times.
- Some solar cooker designs are affected by strong winds, which can slow the cooking process, cool the food, and disturb the reflector.

5. Carbon Dioxide Reduction

Numerous scientific studies reveal that the level of CO_2 in the atmosphere has increased by 31% over the last 200 years (Panwar N. L. et al., 2010). Aproximately 20 Gt of carbon dioxide have been released into the environment since 1800, solely due to deforestation, which leads to the concentration of methane gas in the atmosphere, responsible for the thickening of the ozone layer. Panwar et al. [88] indicated in their review article that over the period 1971-1995, CO_2 emissions increased at an average rate of 1.7% per year. The scenarios predict a faster growth rate of CO_2 emissions of up to 2.2% per year for the period up to 2020.

In 1996, Nandwani (Shukla S. K., 2009) conducted a study on the environmental benefits of solar cookers in Costa Rica and worldwide, comparing the advantages and limitations of solar cookers with conventional wood-fired cookers and electric cookers. He found that the payback time of a box-type cooker, assuming 6-8 months of use per year, is approximately 12-14 months. Approximately 16.8 million tons of wood can be saved each year and the emission of 38.4 million tons of CO_2 into the atmosphere per year can be prevented by using solar cookers. Furthermore, studies were conducted by Nahar over several years on different models of solar cookers in Indian climatic conditions, with an assessment of the potential for CO_2 emissions. It has been estimated that the payback time varies between 1.28 and 4.82 years depending on the energy carrier (cooking fuel) to be replaced. For the various experienced cookers, the energy saved is estimated at 5175 MJ per year.

Renewable energy resources will play an important role in the future of the planet.

The development of solar cooking systems will help meet cooking energy needs and resolve certain problems associated with traditional cooking methods, particularly in developing countries.

III. WORKING PRINCIPLE AND BASIC MODELS

1. Working Principle of Solar Cookers

1.1. Incident Rays on the Solar Cooker

The incident rays on a solar cooker depend on the geometry of this cooker and the position of the sun. The sun's rays must be parallel to the concentrator's optical axis to be concentrated. The sun's position relative to the solar oven depends on the time of day and the day of the year. To calculate the rays incident on the solar oven, it is possible to use the sun's azimuth and zenith tools for a given location to obtain the angle of incidence on the glass.

- **Azimuth tools**: Azimuth is the horizontal angle between geographic north and the projection of the object onto the horizontal plane.

- **Zenith tools**: The zenith is the point in the sky directly above the observer's head. The nadir

is the point opposite the zenith, located directly below the observer. The incident ray is therefore generally the sun's ray striking the surface of the oven. To calculate this angle, you need to know:

- The Sun's declination δ,
- The obliquity of the ecliptic ε,
- The Sun's ecliptic longitude λ,
- And the Sun's altitude at the meridian h.

The following formula (1) is used to calculate the angle α of incidence of the sun's rays on the surface of the solar cooker as follows:

$$\alpha = \arcsin\left(\frac{\sin(\delta)\cos(\varepsilon) - \cos(\delta)\sin(\varepsilon)\cos(\lambda)}{\cos(h)}\right) \tag{1}$$

The resulting value shows that the sun's ray forms an angle α with the normal to the window. In order to optimize the solar oven's efficiency, the window should therefore be tilted so that this angle is as close as possible to zero. The figure 7 below shows the different angular parameters used:

Figure 7. Different Angular parameters.

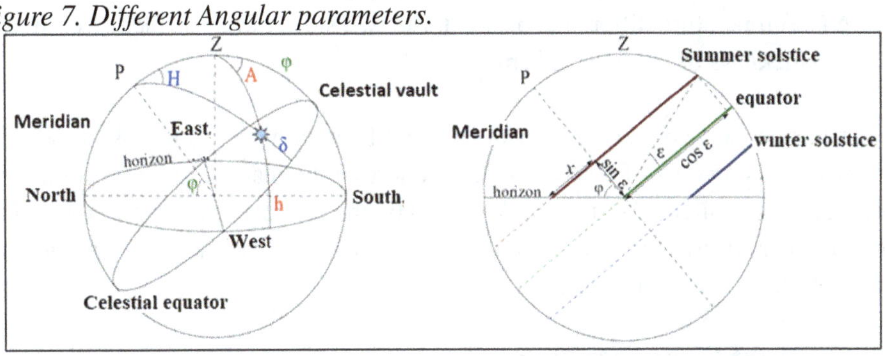

1.2. Heat Transfer Mechanism

Heat transfer is a mode of internal energy exchange between two systems, which can be done in three different ways:

- By conduction,
- By convection,
- By radiation.

Heat transfer in a box-type solar cooker occurs primarily through radiation. The operating principle of a solar cooker is to concentrate the sun's rays on a reflective surface to heat the interior of the box. The heat is then transferred to the food to be cooked by radiation. The temperature inside the box depends on the surface area of the box, the outside temperature, the concentration of the sun's rays, and the insulation quality of the box. The calculation of the heat transfer principle of a box-type solar cooker can be given by the Stefan-Boltzmann law, which defines the relationship between thermal radiation and the temperature of an object considered as a black body (Nahar N. M., 1998).

Heat flux, which reflects the rate of energy transfer through a wall, can be calculated using Fourier's law for thermal conduction using the formula 2:

$$\Phi_t = -k \times A \times \frac{\Delta T}{d} \tag{2}$$

Where:

- Φ_t: Heat flux (in W),
- k: Thermal conductivity of the material (W/m.K),
- A: Surface area through which heat propagates (m^2),
- ΔT: Temperature difference between the two sides of the surface (Kelvin),
- d: Thickness of the material (m).

A solar cooker works on the principle that heat from the sun cooks food in a saucepan. This process of converting light energy into heat energy warms the pot. Here, sunlight is directed onto a cooking pan by the concave mirror. After entering a solar cooker, ultraviolet (UV) beams are converted, in numerous ways, into longer infrared light rays that are impenetrable.

1.3. Sunlight Concentration

A high-spherical reflection mirror surface is used to focus and redirect sunlight into a small cooking area. Even metal and salt can be melted by the strength of the sun, but home solar cooking does not need such high temperatures. Therefore, in the market solar cookers are designed to reach temperatures ranging from 150 °C to 300°C (Mbodji N. and Hajji A., 2017).

1.3.1. Transforming Photon Energy Into Thermal Energy

Concentrated sunlight is directed at a recipient, like a cooking pan. To maximize absorption, solar cooker pots and pans should be matte black. Light energy interacts with the receiver and condenses into heat (Shukla S. K., 2009).

1.3.2. Trapping Heat Energy

By covering the pot with a glass lid, you can boost the cooker's heat-holding capacity, minimize convective energy loss, and maximize light absorption from the top of the pan. Although it is opaque to incoming sunlight, the glazing permits the passage of infrared thermal rays (Yettou F. et al., 2012). Any material that receives sunshine will have some heat conversion. This substance is known as the absorber. The amount that is converted to heat increases with material's color and reflectivity. Therefore, when it comes to converting visible sunlight, black is ideal. Because the reflection is by definition sunlight that has not been converted to heat, a flat material is preferable than one that is highly reflective. Not visible light, or infrared light, is produced by a substance interacting with visible light; it has nothing to do with color according to definition. But certain substances absorb infrared light more effectively than others. But that's usually not a concern when using solar cooking. The angle at which light strikes and reflects off a surface is equal. When creating reflectors, keep this in mind. It is possible for sunlight to strike one of your reflectors, bounce off another reflector farther within your cooker, and then bounce back out again without striking your cooking pot. It is crucial to comprehend how sunlight reflects for this reason. If your reflectors are made correctly, all you have to do to be safe is point them toward the sun.

2. Basic Models

2.1. Box Solar Cookers

This is the most commonly used type in households as depicted in figure 8 below. It consists of a box lined with reflectors and a transparent cover on top. The radiation penetrates the box and is reflected onto a container whose surface absorbs the rays [4]. The transparent cover insulates the interior of the oven from the exterior and traps the radiation to retain as much heat as possible (greenhouse effect).

This cooker can reach temperatures of 100°C to 200°C, depending primarily on the number and size of the reflector panels used. It allows for the slow, even cooking of large quantities of food (Djebara A. and Guettaf Z., 2018). In India, several hundred thousand homes are equipped with this type of cooker.

Figure 8. Solar box cooker.

2.2. Parabolic Solar Cookers

Also called curved concentrator cookers as montred in figure 9 below. They are based on the principle of concentrating light rays (Mullick S. et al., 1996). The light beam is reflected off the parabolic mirror, and the rays converge at single point: the focus of the parabola. This type allows very high temperatures to be reached quickly (Funk P., 2000). Its power depends heavily on the surface area of the parabolic reflector.

Despite its good performance, this cooker requires continuous direct sunlight and must be constantly facing the sun (Shukla S. K., 2009). They are widely used in several hundred thousand homes in Western China.

Figure 9. Parabolic solar cooker.

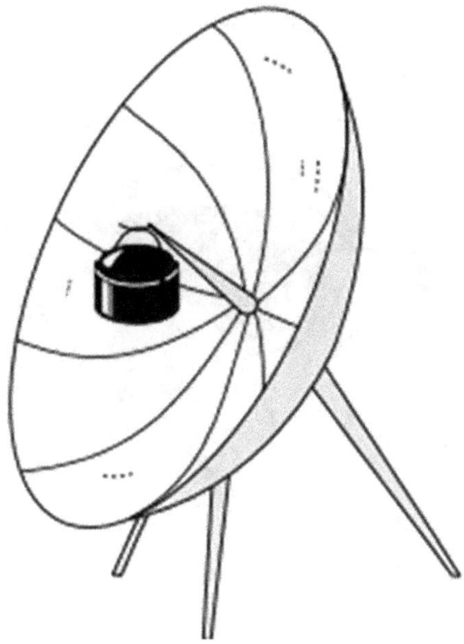

2.3. Panel Solar Cookers

This type of solar cooker combines the two technologies mentioned above (See figure 10). It combines the components of the box cooker and those of the parabolic cooker. It uses both the greenhouse effect principle of the box cooker and the benefits of the parabolic reflector of the parabolic cooker (Funk P., 2000, GAMA A. et al. 2017).

It consists of different panels made from pieces of cardboard covered with reflective paper (aluminum). These panels roughly concentrate the sun's rays onto a pot placed under a glass or plastic bowl to create a greenhouse effect (Panwar N. et al.2012).

It can be used to cook small quantities of food or to clean water by heating it above 65°C. It is less expensive and simple to install due to its use of inexpensive materials. The Cookit brand remains the most widely used in the world.

Figure 10. Solar cooker with panels.

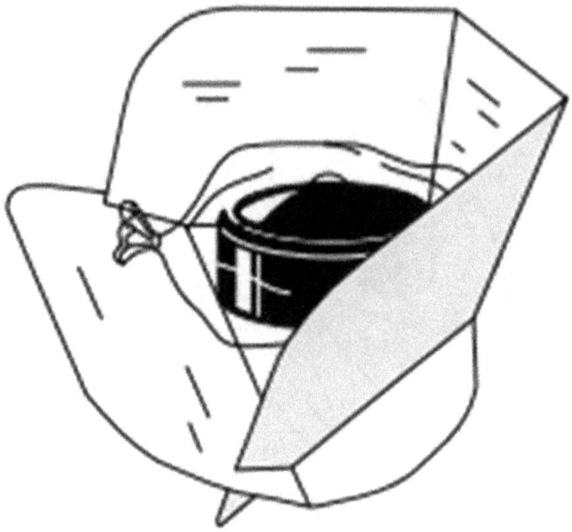

2.4. Indirect Solar Cookers

In this type of solar cooker, the cooking vessel is located outside the solar collector. A circuit is used to transfer heat to the vessel. The main indirect solar cookers are collector-plane solar cookers, vacuum tube solar cookers, and concentrator solar cookers (Djebara A. and Guettaf Z., 2018) (Mbodji N. and Hajji A., 2017). The figure 11 below shows these three types.

Figure 11. Indirect solar cookers: a) with flat sensor, b) with parabolic concentrator and c) with spherical concentrator.

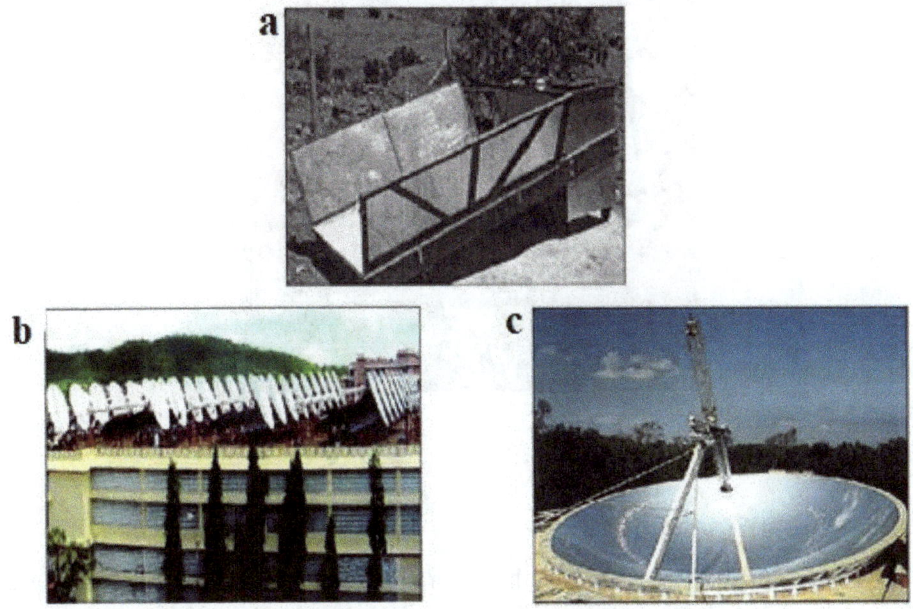

2.4.1. Solar cookers with heat storage

In situations where there is a mismatch between energy demand and supply, thermal energy storage becomes necessary (Patel N. and Philip S., 2000). This type of technology uses a heat-accumulating material to store thermal energy. The stored energy will be used for cooking purposes during unfavorable lighting conditions (clouds, strong winds, etc.) or outside of peak hours (Nahar N., 2003).

Often, thermal energy storage, in the form of a change in the material's internal energy, occurs in three forms: sensible, latent, and thermochemical heat, or a combination of these forms (Nandwani S. S., 1996).

a. Sensible Heat Storage

In this indirect cooking technology, thermal energy is stored by increasing the temperature of a liquid or solid. The figure 12 below illustrates an exemple of this technology.

Figure 12. Storage by sensible heat with engine oil.

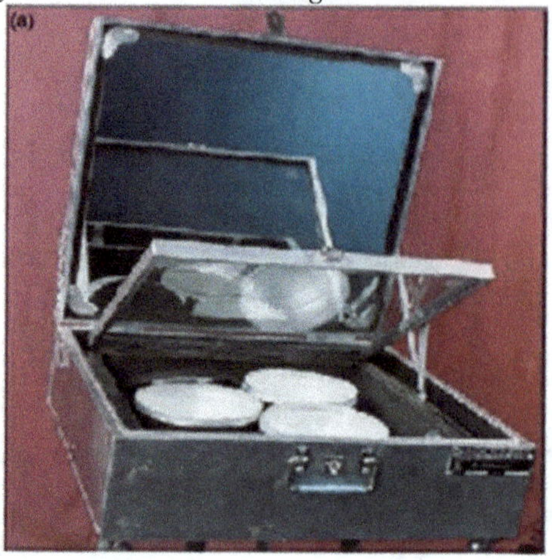

b. Latent Heat Storage

In this type of solar cooker, latent heat storage uses energy stored during a phase change of a material or substance (Mosbah and A. Achour A., 2020).

2.4.2. Fresnel-Lens Solar Cookers

This kind of solar cookers utilize a sequence of inclined steps on the front surface of an otherwise flat sheet of materials like for example glass or plastic in order to to reduce the bulk of a convex concentrating lens (Djeghab A. and Bougoffa E., 2022). The steps are slanted so that light coming through the lens is basically localised on the cooking pot as shown in figure 13. A larger-scale implementation of this idea is made by rerouting sunlight onto the cook pot using a set of reflective concentric rings. If a parabolic reflector were divided into horizontal parts and stacked concentrically, the individual ring angles would essentially be the same. But this would result in shadowing; thus, in order to let light into the cookpot, each ring's diameter needs to be somewhat larger (Yettou F., 2015).

Figure 13. Fresnel-Lens solar cooker: a) Twin Fresnel cooker, and b) larger-scale with reflective concentric rings.

a

b

2.4.3. Photovolaic PV solar cookers

Larger photovoltaic systems powering conventional electrical cooking appliances, but rather smaller off-grid applications. These tiny devices offer an additional method of using sunlight to cook meals (Kassmi K., 2021). Rather than cooking with just direct sunlight, they produce electricity through photovoltaics, which is usually used to power induction stoves. Batteries are typically used to store energy for later use as you can see in figure 14 below:

Figure 14. Photovoltaic cooker system.

IV. CHARACTERIZATION AND EVALUATION PARAMETERS OF SOLAR COOKERS

1. Reached Temperatures and Cooking Times Parameters

The operating principle of a solar tracker is based on the conversion of solar rays into heat through an absorber. The latter generally has a dark surface to cook food over time. In some cases, solar energy is multiplied by the addition of mirrors. For efficiency, heat must be retained as much as possible (Ashok K . and Sudhir C.V., 2009).

In box cookers, a pane of glass is used to keep the heat inside the box through a greenhouse effect. In addition, adding a conventional insulator to the other walls improves heat retention.

For parabolic cookers, mirrors are used to deflect the sun's rays and concentrate them toward the cooker's core (Harmin A. et al., 2007).

The cooking temperature of a solar cooker is reached when solar energy gains equal heat losses to the outside. This is considered a balance between heat gains and heat losses.

The temperature reached by box and panel solar cookers depends primarily on the number and size of reflectors used (Pandey A. K. et al., 2012).

Box cookers and those with reflective panels have a temperature range of 100 to 200°C.

The temperature reached by other parabolic cookers and their derivatives varies between 150 and 250°C, depending on the reflector diameteras as illustrated in figure 15 in next page.

Cooking times vary greatly. They depend heavily on various factors, including the amount of sunlight, the time of year, the time of day, wind strength, the type of container used, and the quantity of food being prepared (Khennoufa A. and Lachibi O., 2017).

Dark surfaces absorb sunlight better than light surfaces. This is why food cooks better in dark, shallow metal containers with suitable lids to retain heat and moisture (Pandey A. K. et al., 2012). Cooking in a solar cooker with a reflector box takes twice as long as in a traditional one. This is without burning the food. Furthermore, the user doesn't need to monitor or stir the food. It can be removed from the cooker later while still hot (Pandey A. K. et al., 2012). In panel cookers, small portions are often cooked in a single container, which speeds up cooking (Pandey A. K. et al., 2012).

Cooking in a parabolic cooker is very similar to cooking over a regular stovetop. Since the concentrated sunlight reaches the bottom of the container, it heats up and cooks very quickly. However, careful stirring and monitoring are required, and even slowing down cooking can slow things down if the dish is oriented in a less-than-optimal manner (Harmin A. et al., 2007).

Figure 15. Box and parbolic solars cooking temperatures.

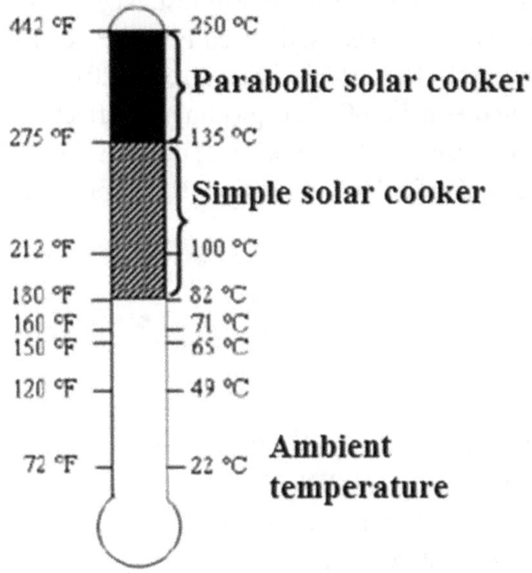

The sun must be high enough in the sky to allow for cooking. Also, there must be no clouds or wind. In this case, the length of the cooker's shadow is shorter than its actual size. Therefore, it is ideal to cook between 9 a.m. and 3 p.m.

Dark-colored containers are used for cooking to better absorb heat, while light-colored ones reflect sunlight.

2. Performance Evaluation Parameters of Solar Cookers

Evaluating the thermal performance of solar cookers is essential for characterizing these devices. Mullick et al. and Funk developed thermal test procedures for both box and parabolic cookers (Djebara A. and Guettaf Z., 2018, Patel N. and Philip S., 2000).

2.1. Evaluation According to Mullick et al. Theory

2.1.1. Box Solar Cookers

According to the IS13429:2000 standard, two main tests are performed. These tests are a stagnation test and a sensible heat test. These two tests determine the two thermal performance parameters, denoted F_1 and F_2, called parameters or figures of merits (Panwar N. et al.2012, GAMA A. et al. 2017). These parameters are determined based on thermal profiles, neglecting the effect of reflecting mirrors.

F_1 is defined as a measure of the temperature difference received by the absorber plate at a particular stage of solar illumination. It is expressed as with equation 3:

$$F_1 = \frac{\eta_0}{U_L} = \frac{T_{ps} - T_{as}}{I_s} \tag{3}$$

Where:

- η_0: Optical efficiency of the cooker (%),
- U_L: Heat transfer coefficient (W/m²),
- T_{ps}: Absorber temperature at stagnation (°C),
- T_{as}: Ambient temperature at stagnation (°C),
- I_s: Incident solar irradiance on a horizontal surface at stagnation (W/m²).

F_2 is derived from the sensible heat test. It is defined by the following relationship 4:

$$F_2 = \frac{F_1 \times (m c_p)_w}{A_{sc} \times \tau} \ln\left[1 - \frac{1}{F_1}\left(\frac{T_{wi} - \overline{T}_a}{\overline{I}_s}\right)\right] / \left[1 - \frac{1}{F_1}\left(\frac{T_{wf} - \overline{T}_a}{\overline{I}_s}\right)\right] \tag{4}$$

Where:

- $(mc_p)_w$: Heat capacity of water (J/°K),
- A_{sc}: Solar cooking surface (m²),
- T_{wi}: Initial water temperature (°C)
- T_{wf}: Final water temperature (°C)
- τ: Time interval during which the water temperature increases from T_{wi} to T_{wf} (min),
- $\overline{T}_a$: Average ambient air temperature every 10 minutes (°C),
- $\overline{T}_s$: Average solar illumination received by a horizontal surface (W/m²).

Mullick also defined the time constant τ_{boil} for the sensible heat phase from room temperature to 100°C by the relation 5:

$$\tau_{boil} = \frac{-F \times_1 (mc_p)_w}{F_2 \times A_{sc}} \ln\left[1 - \frac{1}{F_1}\left(\frac{100 - T_a}{\overline{I_s}}\right)\right] \quad (5)$$

Where:

T_a: Ambient temperature (°C).

Funk introduced the Adjusted or Standardized Cooking Power P_s parameter [2]. This term is given by the following relationship 6:

$$P_s = \frac{700 \times (M_w \times C_w) \times \Delta T_w}{600 \times I_b} \quad (6)$$

Where:

M_w: Mass of water (Kg/Kmol),

C_w: Specific heat Capacity of water (J/Kg°K),

ΔT_w: Difference between the final and initial water temperatures (°C),

Tw: Water temperature in the pot (°C),

I_b: Average direct solar radiation every 10 minutes (W/m^2),

I_e: Average solar irradiance (W/m^2),

I_n: Solar irradiance standardized to 700 W/m^2,

This parameter P_s plays an important role; it facilitates comparisons between various solar cooker designs under different irradiance intensities. Cooking power must be evaluated under the following conditions:

- Wind speed less than 1 m/s,
- Water temperature inside the cooking pot must be between 40 and 90°C,
- And ambient temperature and solar irradiance must be between 20 and 35°C and 450 and 1100 W/m^2 respectively.

2.1.2. Parabolic Solar Cookers

For this type of cooker (Yakhlef Y. and Mehdaoui A., 2018), two parameters are used to evaluate their performances: the heat loss factor and the optical efficiency factor (Mullick S. et al., 1996).. The heat loss factor $F'U_L$ depends primarily on the temperature of the water in the cookware, the wind speed, and the surface area of the cookware. It is determined from a cooling test according to the following equation 7:

$$F'U_L = \frac{(MC)'_w}{A_1 \times \tau_0} \quad (7)$$

Where:

A_1: Surface area of the cooking utensil,

$(MC)'_w$: The combined calorific value of the water and the cooking utensil, which is given by the relationship 8 below:

$$(MC)'_w = M_w \times C_w + M_p \times C_p \quad (8)$$

M_w: The mass of water,

C_w: The specific heat capacity of water,

M_p: The mass of the pot,

C_p: The specific heat capacity of the pot (usually aluminum) and the utensil.

The optical efficiency factor $F'\eta_0$ is determined from the heating test. It depends mainly on the amount of reflector surface area, the orientation of the concentrator, the intersection area, and the absorption of the outer surface of the cookware, and the reflectivity of the concentrator surface (Yakhlef Y. and Mehdaoui A., 2018). It is defined as follows with equation 9 below:

$$F'\eta_0 = \frac{F'U_L}{c} \frac{\left[\left(\frac{T_{w2} - \overline{T}_a}{\overline{I}_b} \right) - \left(\frac{T_{w1} - \overline{T}_a}{\overline{I}_b} \right) e^{-\tau/\tau_0} \right]}{(1 - e^{-\tau/\tau_0})} \quad (9)$$

Where:

C: The ratio of the opening area of the parabola Ap to the surface area of the cooking utensil Ai.

T_{w1} and T_{w2}: the initial and final water temperatures, respectively, during the time interval t, which is 10 minutes.

$\overline{T}_a$: The average ambient air temperature every 10 minutes,

$\overline{I}_b$: The average direct solar radiation every 10 minutes.

2.2. Evaluation by Funk

According to the international standard developed by Funk, solar cookers are evaluated by calculating the cooking power and its energy efficiency (Yakhlef Y. and Mehdaoui A., 2018). Cooking power is assessed by recording the temperature of a quantity of water inside the cooking utensil (GAMA A. et al. 2017). The temperature reading is taken every 10 minutes for a period not exceeding 4 hours, without this temperature exceeding 95°C. The average power P delivered during this interval is given by this relationship (10):

$$P = \frac{(M_w \times C_w) \times \Delta T_w}{\Delta t} \quad (10)$$

Where, Δt: Time interval generally taken equal to 10 minutes.

According to the first law of thermodynamics, the energy efficiency of the cooker is defined as the ratio of the output energy E_o to the input energy E_i (Nahar N., 2003).

In the case of a parabolic solar cooker, this efficiency η is given by the equation 11:

$$\eta = \frac{E_0}{E_i} \frac{(MC)_w \times (T_{wf} - T_{wi})}{I_b \times \Delta t \times A_p} \tag{11}$$

Where:

T_{wi} and T_{wf}: the initial and final water temperatures, respectively,

A_p: Opening surface of the parabola (m^2).

3. Empirical Evaluation of the Thermal Performance of Solar Cookers

Characterization is achieved through a series of tests and measurements performed on the considered solar cookers. These tests are conducted over several days under varying climatic conditions and with a carefully chosen adjustment period. During these tests, temperatures and illumination are measured using appropriate equipment. In addition to calculating the evaluation parameters, heating and cooling curves are also plotted and interpreted. This allows for verification of the feasibility, performance, and profitability of the solar cookers tested (Nandwani S. S., 1996).

The position of the sun varies throughout the day. To collect maximum solar radiation, solar cookers must be continuously redirected toward the sun, following its path during the cooking phase (Yettou F. et al. 2012).

Typically, the cooker adjustment time is between 10 and 20 minutes, starting from the initial position (10:00 a.m.) and ending at 3:00 p.m. Temperatures and illumination are measured using measuring instruments, namely thermometer and pyranometer (Nahar N. M., 1998).

Solar cooker characterization can be done in a sophisticated and very simple manner with a data acquisition card and a suitable microcomputer. This card is associated with a set of sensors (temperature sensors, pyranometer, etc.) and a weather station (wind direction, speed, and humidity).

The data collected in real time will be displayed in digital and graphical formats and saved to a file format. They will then be imported into Excel and displayed on the microcomputer screen.

3.1. Water Temperature Evolution Test

Cooking often requires boiling water or heating the cooked items to a high water content. This test is generally conducted on a clear to partly clear day in June (Mullick S. et al., 1996). The parameters to be recorded are: ambient temperature, water temperature in the cooker, direct solar radiation I_b, and solar irradiance standardized to 700 W/m^2 noted I_n.

The curve to be recorded during this test is the measured temperature profiles (water temperature T_w and ambient air temperature T_a) for heating a given quantity of water as a function of time, with variations in solar irradiance during the day corresponding to the test.

3.2. Water Heating Test

For a given water load, which depends on the solar cooker's capacity to characterize, the cooker is placed facing the sun (Funk P., 2000). The curves showing the temperature changes over time and the corresponding radiation changes are also shown, discussed, and interpreted.

3.3. Water Cooling Test

For this test, the solar cooker is not directed toward the sun. It is completely shaded, and the container is hidden. For a given water load, temperature curves over time are plotted. Solar illumination curves are also shown (Mbodji N. and Hajji A., 2017).

If this test is conducted on clear days, solar cookers can reach stagnation without requiring much time. Also, the greater the water load, the longer it takes for the solar cooker to heat the water in its container.

3.4. Cooking Power and Instant Performance

The thermal performances which correspond to the cooking power and the instantaneous yield are evaluated by the graphic technique for a well-defined water load (Khennoufa A. and Lachibi O., 2017).

Cooking power is plotted against temperature difference (between water temperature T_w and ambient air temperature T_a): $\Delta T = T_w - T_a$.

While the instantaneous yield is represented as a function of the ratio: $\Delta T/I_n$ (I_n is the solar irradiance standardized to 700 W/m^2).

The slope of the curve gives the value of the overall heat loss coefficient U_L, and its gradient determines the cooking power of the cooker used. These two characteristics are highly dependent on the water load used to characterize a solar cooker (Panwar N. L. et al., 2011).

The maximum efficiency value is obtained by the intersection with the y-axis.

3.5. Energy and Exergy Evaluation of Solar Cookers

Several theoretical and experimental studies have been conducted in recent years to evaluate the energy and exergy efficiencies of solar cookers around the world. (Panwar et al, 20) experimentally evaluated the energy and exergy efficiency of a solar cooker intended for cooking animal food. The solar cookers are made using cement, bricks, glass lids, wood, and a mild steel absorber plate (REH S., 2004).

A diagram with view of the completed cooker are shown in figure 16. The output energy of this cooker is between 1.89 kJ and 49.4 kJ, while the output exergy is between 0.11kJ and 2.72 kJ during the same time interval. The cooking appliance's energy efficiency ranged from 1.12% to 29.78%, while its exergetic efficiency ranged from 0.07% to 1.52% during this test period (Panwar N. L. et al., 2011).

Figure 16. Solar cooker for cooking animal food: a) dimensions, b) side view.

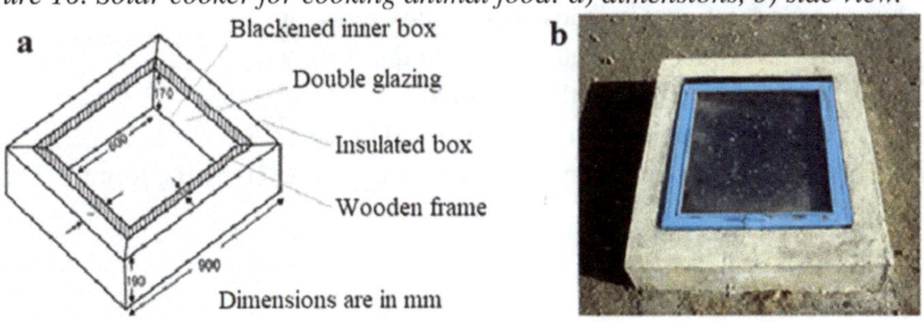

At the end of 2009, Shukla (Yettou F., 2015) compared the energy and exergy yields of a parabolic cooker for domestic use and another for collective use. According to the test results, it was observed that the output power of the collective cooker ranges between 2.73W and 43.3 W, and that of the domestic cooker ranges between 7.77W and 33.4W. The output energy of the collective cooker is in the order of 1.92W to 2.58W, while for the domestic cooker, it varies from 0.65W to 1.45W. The energy efficiency of a collective parabolic cooker varies between 8.3% and 10.5%, and for a domestic parabolic cooker, it ranges from 7.1% to 14.0%.

In early 2011, Pandey et al. presented a comparative experimental study between a box-type solar cooker and a parabolic-type cooker based on exergy analysis. The experiments were carried out with the cookers filled with different amounts of water and rice. The results showed that the energy efficiency of the cookers is proportional to the volume of water used. They also found that the exergetic efficiency of cookers varies with the nature of the food being cooked. However, the exergetic efficiency of the parabolic cooker is always higher than that of the box solar cooker.

3.6. Heat Losses Calculation

Convection and radiation are the absorber's two main sources of heat loss (Nahar N. M., 1998). The total heat loss from the absorber is the total heat loss from the vessel's top and the sides of the thermal tank (Reh S., 2004). The amount of heat lost through radiation and convection is given by these relationships 12 and 13 below:

$$Q_{conv} = Q_{conv,side} + Q_{conv,top} \tag{12}$$

$$Q_{rad} = Q_{rad,side} + Q_{rad,top} \tag{13}$$

Where:
Q_{conv} = Heat loss from the convection-absorbed,
$Q_{conv,side}$ = Convectional heat loss from the thermal tank side,
$Q_{conv,top}$ = Convectional heat loss from the vessel's top,
Q_{rad} = Heat loss caused by radiation from the absorber,
$Q_{rad,side}$ = Heat loss from the thermal tank side by radiation,
$Q_{rad,top}$ = Heat loss from the top of the vessel by radiation.
Therefore, total heat loss from the absorber Q_{loss} is given by the formula 14:

$$Q_{loss} = Q_{conv} + Q_{rad} \tag{14}$$

The following equation determines the overall thermal efficiency $\eta_{overall}$ using the relationship 15:

$$\eta_{overall} = {M_f \times C_{pf} \times \Delta T_f} \Big/ {I_s \times A_{sc} \times \Delta t} \tag{15}$$

Where:
M_f: Mass of cooked food (kg),
C_{pf}: Specific heat capacity of food (kJ/kg.k),
ΔT_f: Change in food temperature (°C),
I_s: The cooker aperture being struck by insolation (W/m^2),
A_{sc}: Aperture area of collector (m^2),
Δt: The entire amount of time needed to cook food.

V. CONCLUSION

Solar energy has several advantages. It allows electricity to be produced without raw materials and without impacting the environment. It is currently used to meet a wide range of needs: household appliances, heating, boilers, lighting, etc.

Solar cooking, which emerged in the late 1960, involves preparing meals using a solar oven, also known as a solar cooker. Small solar cookers can reach temperatures of up to 150°C. Parabolic cookers allow the same dishes to be prepared using a gas or electric stove.

Using solar energy for cooking food helps reduce deforestation in some countries where cooking with coal and wood is the norm. It also reduces CO_2 emissions into the atmosphere.

As part of this work, we conducted a literature review and a state-of-the-art analysis of solar cookers, their development, and their use worldwide. We presented the essential parameters for evaluating thermal performance. These parameters are essential for conducting tests to gain insight into the thermal behavior and efficiency, as well as the feasibility of cooking various foods at relatively high temperatures.

We are currently working on improving the thermal performance of a commercial concentrated solar cooker by optimizing efficiency due to low temperatures by replacing fixed mirrors with movable mirrors with specific joints.

As a result of this study, we have found that the use of solar cookers can result in energy savings and environmental friendliness. Therefore, it seems important for our country to begin mastering this technology to benefit from these advantages in the future and ensure a certain balance in energy consumption by exploiting all renewable energy resources.

REFERENCES

Ashok, K. & Sudhir, C.V. (2009). Proposal for new world standard for testing solar cookers. *Journal of Engineering Science and Technology*. Vol. 4, N°. 3. pp.272 – 281. School of Engineering, Taylor's University College. India.

Bouhamida, M., & Henichi, I. (2020). *Etude et conception d'un cuiseur solaire à réflecteur composé avec tube sous vide*. Université de Ghardaïa.

Djebara A. and Guettaf Z. (2018). Etude comparative entre deux cuiseurs solaires testés sous conditions climatiques de Ghardaïa. 2018-2019.

Djeghab A. and Bougoffa E. (2022). Etude expérimentale d'un cuiseur solaire à faible cout pour les applications domestiques, Master Sciences et Techniques, Spécialité Energie renouvelables en Mécanique, 2022-2023.

Funk, P. (2000). Evaluation the international standard procedures for testing solar cookers and reporting performance. *Solar Energy*, *68*(11), 1–7.

Gama, , A. et al. (2017). Détermination des performances d'un cuiseur solaire boite suivant les procédures de Mullik et Funk. *Rev. Roum. Sci. Techn.* – [Bucarest.]. *Electrotechnique et Energie*, *62*(2), 157–164.

Harmin, A., Boukar, M., & Amar, M. (2007). Etude expérimentale d'un cuiseur solaire de type boite à trois reflecteurs plans et une surface d'ouverture inclinée. *Revue des Énergies Renouvelables*, *10*(11), 31–38.

Harmin, A., Boukar, M., & Amar, M. (2007). Augmentation du gain énergétique d'un cuiseur solaire à double exposition. *Colloque International sur les Energies Renouvelables CER07, 4-5 Mai 2007*, Oujda. Maroc.

Kassmi, K. (2021). Autonomous solar cooker with photovoltaic energy. *New Journal of Science Vulgarization.*, *1*(2), 54–62.

Khennoufa A., Lachibi O. (2017). *Etude d'un cuiseur solaire*. 2017-2018.

Mbodji, N., & Hajji, A. (2017). Analyse technico-économique et évaluation de l'impact environnemental de la cuisson solaire directe au Maroc. Rev. Mar. sci. Agron. *Vét.*, *5*(3), 325–332.

Mosbah, A. & Achour, A. (2020). Contribution à la détermination expérimentale des performances d'un cuiseur solaire testé sous climat aride du Sude Algérien, master Energies Renouvelables en Electrotechnique, Université de Ghardaia, Faculté des Sciences et Technologies Département d'Automatique et Electromécaniques, 2020-2021.

Mullick, S., Kumar, T., & Kumar, S. (1996). Testing of box type solar cooker: Second figure of merit F2 and its variation with load and number of pots. *Solar Energy*, *57*(15), 409–413.

Nahar, N. (2003). Performances and testing of a hot box storage solar cooker. *Energy Conversion and Management*, *44*, 1323–1331.

Nahar, N. M. (1998). Design, development and testing of a novel non-tracking solar cooker. *International Journal of Energy Research*, *22*, 1191–1198.

Nandwani, S. S. (1996). Solar cookers – cheap technology with high ecological benefits. *Ecological Economics*, *17*, 73–81.

Pandey, A. K., Tyagi, V. V., Park, S. R., & Tyagi, S. K. (2012). Comparative experimental study of solar cookers using exergy analysis. *Journal of Thermal Analysis and Calorimetry*, *109*, 425–431.

Pandey, A. K., Tyagi, V. V., Park, S. R., & Tyagi, S. K. (2012). Comparative experimental study of solar cookers using exergy analysis. *Journal of Thermal Analysis and Calorimetry*, *109*, 425–431.

Panwar, N., Kaushik, S., & Kothari, S. (2012). State of the art of solar cooking: An overview. *Renewable & Sustainable Energy Reviews*, *16*(6), 3776–3785.

Panwar, N. L., Kaushik, S. C., & Kothari, S. (2011). Role of renewable energy sources in environmental protection: A review. *Renewable & Sustainable Energy Reviews*, *15*, 1513–1524.

Panwar, N. L., Kothari, S., & Kaushik, S. C. (2010). Experimental investigation of energy and exergy efficiency of masonry-type solar cooker for animal feed. *International Journal of Sustainable Energy*, *29*(13), 178–184.

Patel, N., & Philip, S. (2000). Performance evaluation of three solar concentrating cookers. *Renewable Energy*, •••, 347–355.

REH S. (2004). Renewable energy: A response to climate change. *Solar Energy*, *76*, 9–17.

Shukla, S. K. (2009). Comparison of energy and exergy efficiency of community and domestic type parabolic solar cookers. *International Journal of Green Energy*, *6*, 437–449.

Yakhlef Y. and Mehdaoui A. (2018). Contribution à la détermination expérimentale des performances thermiques d'un cuiseur solaire parabolique. 2018-2019.

Yettou, F. (2015). Conception et réalisation d'un système de cuisson solaire destiné au site saharien (Ghardaïa, Algérie). Thèse de Doctorat en Sciences. Université Hadji Lakhdar Batna Faculté de Technologie Département d'Electrotechnique. Soutenue le 10 Mars 2015.

Yettou F. et al. (2012). Etude comparative entre deux types de cuiseurs solaires boites en milieu aride. *Revue des Energies Renouvelables*, Vol. Numéro spécial Sienr'12, pp. 233-245.

Chapter 8
Health and Safety Implications of Solar Cooking

C. Selvamurugan
https://orcid.org/0000-0003-3447-1970
Dhaanish Ahmed Institute of Technology, India

P. K. Nishma
Dhaanish Ahmed Institute of Technology, India

P. R. Delna
Nehru Arts and Science College, India

Hisha Sherin K.
Dhaanish Ahmed Institute of Technology, India

Fathimath Jaseela
Dhaanish Ahmed Institute of Technology, India

Tharisha Nourin
Dhaanish Ahmed Institute of Technology, India

ABSTRACT

Solar cooking is emerging as a sustainable and health-conscious alternative to traditional cooking methods, particularly in regions facing energy crises. This review explores the health and safety implications of solar cooking by addressing its advantages over conventional fuel-based cooking. Traditional cooking methods, especially those relying on biomass and fossil fuels, pose severe health risks, including respiratory diseases due to indoor air pollution. Solar cooking eliminates smoke emissions, significantly reducing the incidence of lung infections, eye irritation, and other health issues. Recent innovations in solar cooker design have improved both performance and safety, making these devices more user-friendly and accessible. By evaluating solar cooking's impact on health and safety, this review highlights its potential as a viable solution for sustainable cooking in the post-energy crisis era.

DOI: 10.4018/979-8-3373-2535-4.ch008

1. INTRODUCTION

The global shift toward sustainable energy solutions has spurred interest in solar cooking, a method that harnesses solar energy to prepare food. As concerns over environmental degradation, rising fuel costs, and health hazards associated with traditional cooking methods grow, solar cooking is increasingly being viewed as a viable alternative. Solar cookers use reflective panels, insulated containers, and heat-absorbing materials to concentrate sunlight and convert it into thermal energy for cooking. This process eliminates the need for conventional fuels like firewood, charcoal, and liquefied petroleum gas (LPG), making it a cleaner and more sustainable option. Traditional cooking methods, particularly those involving biomass fuels, pose significant health risks, especially in developing countries where indoor cooking with open fires or inefficient stoves is common. According to the World Health Organization (WHO), household air pollution from cooking with solid fuels contributes to millions of premature deaths annually, primarily due to respiratory infections, cardiovascular diseases, and chronic obstructive pulmonary disease (COPD). Women and children, who spend the most time near cooking fires, are disproportionately affected. Solar cooking eliminates indoor air pollution, offering a safer and healthier way to prepare meals. Beyond health benefits, solar cooking plays a crucial role in environmental conservation. Deforestation is a major concern in regions where wood is the primary cooking fuel. The use of solar cookers can significantly reduce the demand for firewood, helping to preserve forests and mitigate climate change. Additionally, solar cooking reduces carbon emissions associated with burning fossil fuels, contributing to global efforts to combat environmental degradation.

The growing popularity of solar cooking is also driven by advancements in solar technology. Early solar cookers were often inefficient and impractical for everyday use. However, modern designs have improved significantly, incorporating better insulation, heat-retentive materials, and adjustable reflectors to enhance efficiency. Portable and hybrid solar cookers now allow for greater flexibility, enabling cooking even in less-than-ideal sunlight conditions. These innovations have expanded the appeal of solar cooking beyond rural communities to urban areas and emergency relief settings.Despite these advantages, solar cooking is not without its challenges. The primary limitation is its dependence on sunlight, which restricts cooking during cloudy days or at night. While thermal storage systems and hybrid solar-electric models are being developed to address this issue, widespread adoption remains hindered by factors such as initial costs, lack of awareness, and cultural preferences for traditional cooking methods. Additionally, safety concerns such as burns from concentrated heat and the need for proper handling of solar cookers must be addressed to ensure safe and effective usage.This review examines the health and safety implications of solar cooking, focusing on its benefits, safety features, and potential risks. It highlights how solar cooking reduces indoor air pollution, lowers health risks, and contributes to environmental sustainability. Furthermore, it explores safety precautions, design innovations, and the challenges that must be overcome for solar cooking to become a mainstream solution. As the world seeks sustainable and health-conscious cooking methods, solar cooking presents a promising alternative that aligns with both public health and environmental goals.

The review was written employing a narrative approach. Applicable literature was gathered from databases like ScienceDirect, PubMed, SpringerLink, and Google Scholar by employing keywords like solar cooking, health implications, food safety, safety hazards, clean cooking technologies, and environmental sustainability. Reports from worldwide organizations like the World Health Organization (WHO), Food and Agriculture Organization (FAO), and United Nations Environment Program (UNEP) were also reviewed. Only peer-reviewed English-language articles from English language journals, technical reports, and review papers between 2000 and 2025 were reviewed. Research on the health effects, safety concerns, hygienic factors, and environmental consequences of solar cooking was given top priority, while extraneous papers were screened out. The chosen studies were critically appraised, contrasted, and synthesized to determine existing knowledge, practical applications, and knowledge gaps within the realm of solar cooking.

2. INTRODUCTION TO SOLAR COOKING AND ITS GROWING POPULARITY

Solar cooking is an innovative, sustainable, and energy-efficient method of food preparation that utilizes the sun's energy to generate heat. As concerns about climate change, deforestation, and indoor air pollution increase, solar cooking has gained recognition as a viable alternative to traditional cooking methods that rely on firewood, charcoal, or fossil fuels. This method not only reduces environmental degradation but also provides significant health and economic benefits to communities worldwide. While solar cooking has been around for centuries, advancements in technology, materials, and design have significantly improved its efficiency and accessibility. Today, various solar cookers are available, ranging from simple homemade devices to sophisticated commercial models. This growing popularity is driven by the global push toward renewable energy, the need for sustainable cooking solutions, and the rising cost of conventional fuels. This section explores the working principles of solar cooking, its historical evolution, its benefits, and the factors contributing to its increasing adoption. (Panwar N. L et al., 2012)

- **How Solar Cooking Works**

Solar cooking harnesses sunlight and converts it into heat energy for cooking food. Unlike conventional stoves that burn fuel, solar cookers use reflectors, heat-absorbing surfaces, and insulating materials to trap and retain heat. The fundamental principle behind solar cooking is the greenhouse effect, where solar radiation is captured and retained within a confined space, raising the temperature high enough to cook food efficiently. (Funk & Larson, 1998)

- **Most solar cookers operate through a combination of three main mechanisms:**

1. **Solar Concentration –** Reflective surfaces such as mirrors or aluminum foil are used to direct sunlight onto a focal point, increasing heat intensity. Parabolic and panel cookers rely heavily on this principle to achieve high cooking temperatures.
2. **Heat Absorption –** Cooking vessels are typically painted black or made from dark materials that efficiently absorb and convert solar radiation into heat.
3. **Heat Retention –** Transparent lids, glass covers, or insulation materials trap heat inside the cooking chamber, preventing heat loss and maintaining a stable cooking temperature.

Solar cookers can reach temperatures ranging from 80°C to 250°C (176°F to 482°F), depending on the cooker type and sunlight availability. While cooking times vary, most meals can be prepared within 1 to 3 hours, making solar cookers suitable for slow-cooked dishes, boiling, steaming, and baking. (Ana C et al., 2025)

Table 1. Intensities are approximate average and can vary by season, latitude and weather.

Time of Day	**Approximate Solar Light Intensity (W/m^2)**
Morning (8:00-10:00 AM)	300 – 500
Midday / Noon (11:00 AM – 2:00 PM)	800 – 1000
Afternoon (3:00 – 5:00 PM)	400 – 600
Evening (after 5:00 PM)	< 200
Night	0

The concept of using the sun's energy for cooking has been around for centuries, but it has evolved significantly over time. Early civilizations, particularly in Egypt and Mesopotamia, utilized polished metal surfaces to concentrate sunlight and generate heat. However, structured solar cooking devices did not emerge until much later. During the 19th and early 20th centuries, scientific interest in solar cooking grew, with researchers developing various models aimed at improving efficiency. In 1878, French engineer Augustin Mouchot demonstrated a solar concentrator that could generate steam for cooking and other applications. His invention showcased the potential of solar energy in reducing dependence on fossil fuels. However, with the advent of coal and later electricity, interest in solar cooking declined until the energy crises of the 1970s reignited the need for sustainable alternatives.

Modern solar cooking technologies have benefited from advancements in materials science and thermal engineering. Today, solar cookers come in different designs, including box cookers, parabolic cookers, and panel cookers. Box cookers use insulated chambers and transparent covers to trap heat, making them ideal for slow cooking. Parabolic cookers, which focus sunlight onto a single point, can achieve higher temperatures suitable for frying and boiling. Panel cookers combine reflective surfaces with insulated containers to maximize heat retention while remaining affordable and easy to construct. (Kimambo, 2007)

Figure 1. Development Timeline of Solar Cookers

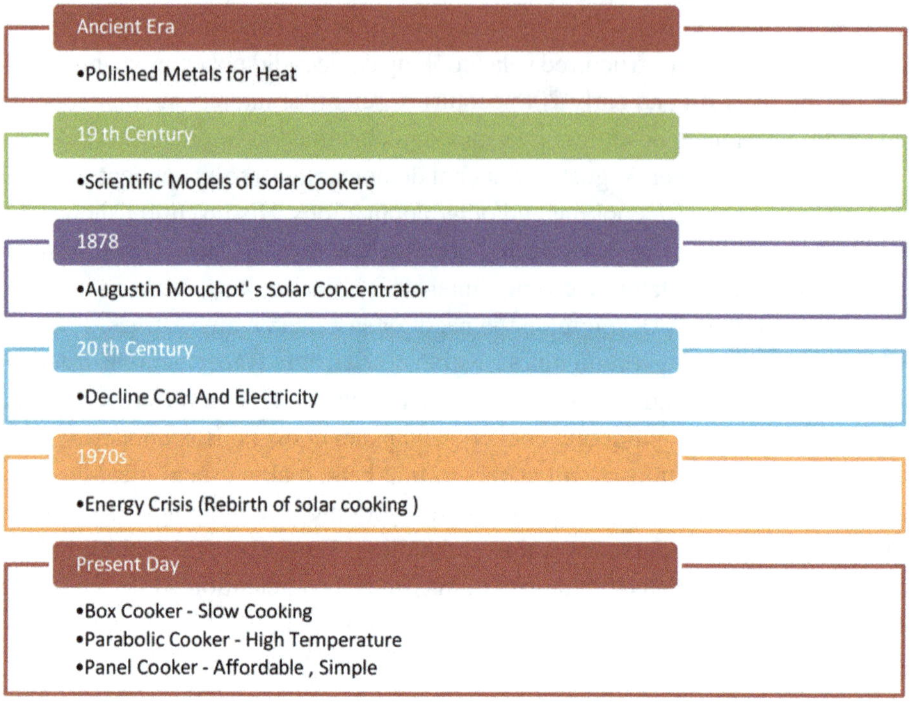

- **Benefits Driving the Popularity of Solar Cooking**

The growing popularity of solar cooking is driven by its numerous benefits across health, environmental, economic, and social aspects. Health-wise, it eliminates exposure to harmful indoor air pollution caused by traditional stoves, reducing respiratory diseases, eye irritation, and carbon monoxide poisoning. Environmentally, solar cooking helps combat deforestation by reducing reliance on firewood and charcoal while also cutting greenhouse gas emissions and air pollution. Since solar energy is abundant and renewable, it provides a sustainable alternative to fossil fuels. Economically, solar cookers lower household expenses by eliminating the need for costly fuels like LPG, kerosene, and firewood, making them particularly beneficial for low-income communities. Once installed, they require minimal maintenance and have no operational costs, offering long-term financial savings. Socially, solar cooking reduces the burden on women and children, who often spend hours collecting firewood, allowing them more time for education and economic activities. It also enhances food security in off-grid regions by providing a reliable cooking method without dependency on external fuel supplies. Additionally, solar cooking promotes energy independence and sustainability, supporting global efforts toward clean energy solutions. With its ability to provide a smoke-free, cost-effective, and eco-friendly cooking method, solar cooking is emerging as an essential solution for communities worldwide, helping to address key challenges such as air pollution, fuel scarcity, and household energy expenses. As awareness of its benefits grows, more individuals and governments are adopting solar cooking technologies to improve quality of life while protecting the planet. (Biran A et al., 2004)

3. HEALTH RISKS OF TRADITIONAL COOKING METHODS

Traditional cooking methods, which rely on biomass fuels such as firewood, charcoal, crop residues, and dung, as well as fossil fuels like kerosene and liquefied petroleum gas (LPG), are widely used in many parts of the world. While these methods have sustained human societies for centuries, they pose serious health risks to individuals and communities, particularly in low-income and rural areas where cleaner alternatives are unavailable. The health hazards associated with traditional cooking methods stem from indoor air pollution, fire-related injuries, exposure to toxic fumes, and long-term respiratory and cardiovascular diseases. This section explores these risks in detail, highlighting their impact on human health and well-being. (Wangxin Liu et al., 2023)

a. Carbon Monoxide Poisoning and Neurological Effects

Carbon monoxide (CO) is another hazardous byproduct of traditional cooking. Unlike other pollutants, CO is colorless, odorless, and highly toxic, making it difficult to detect. In poorly ventilated spaces, CO levels can rise quickly, leading to acute poisoning that manifests as headaches, dizziness, nausea, confusion, and in severe cases, death due to suffocation.

Long-term exposure to low levels of carbon monoxide can cause neurological damage, affecting memory, concentration, and cognitive function. Studies have also linked CO exposure to an increased risk of stroke, heart disease, and developmental disorders in children. Pregnant women exposed to CO from cooking smoke may experience complications such as low birth weight, stillbirth, and developmental delays in infants. (Ning K et al., 2020)

b. Fire Hazards and Burn Injuries

Traditional cooking methods, particularly open fires and kerosene stoves, pose a significant risk of burn injuries and household fires. In many rural communities, cooking is done on three-stone fires or rudimentary stoves placed at floor level. This setup increases the likelihood of accidental burns, especially among young children who may come into direct contact with flames or hot cooking utensils. (Mehta K et al., 2023)

Kerosene stoves, which are commonly used in low-income households, are highly flammable and prone to explosions. If spilled or handled improperly, kerosene can ignite quickly, causing severe burns and property damage. In densely populated areas, house fires triggered by traditional cooking methods have led to fatalities, homelessness, and loss of livelihoods.

c. Cardiovascular and Metabolic Health Impacts

Emerging research suggests that traditional cooking methods contribute to cardiovascular diseases, hypertension, and metabolic disorders. The chronic inhalation of smoke and toxic fumes leads to systemic inflammation, oxidative stress, and arterial damage, increasing the risk of heart attacks and strokes. Studies have shown that individuals exposed to high levels of biomass smoke have higher blood pressure, reduced lung function, and an elevated risk of heart disease compared to those using cleaner energy sources.

Cooking with high-temperature flames and charring foods over open fires also produces heterocyclic amines (HCAs) and polycyclic aromatic hydrocarbons (PAHs), which are carcinogenic compounds. Frequent consumption of foods cooked under such conditions has been linked to increased risks of stomach, liver, and colorectal cancers. Additionally, women who cook in poorly ventilated kitchens are more likely to develop type 2 diabetes and obesity, possibly due to the metabolic effects of prolonged exposure to air pollution. (Rodríguez-Ayala M et al., 2022)

d. Mental Health and Social Burdens

Beyond physical health risks, traditional cooking methods also contribute to mental health issues and social burdens, particularly for women and children. In many developing regions, women spend several hours each day collecting firewood, carrying heavy loads over long distances. This physically demanding task leads to chronic back pain, musculoskeletal injuries, and exhaustion.

The time spent gathering firewood reduces opportunities for education, employment, and social engagement, perpetuating cycles of poverty and gender inequality. In conflict zones and areas prone to violence, women and girls are also at risk of harassment, assault, and gender-based violence while collecting firewood in remote locations.

Additionally, the constant exposure to smoke and harsh cooking conditions can cause stress, anxiety, and depression, particularly among women who bear the responsibility of feeding their families under challenging circumstances. The financial burden of purchasing cooking fuels also contributes to economic stress, forcing families to make difficult trade-offs between buying food, medicine, or fuel. (Farmer, 2021)

Figure 2. Development Timeline of Solar Cookers

- **The Need for Safer and Healthier Cooking Alternatives**

Given the extensive health risks associated with traditional cooking methods, there is an urgent need to transition toward cleaner, safer, and more sustainable cooking solutions. Solar cooking presents a viable alternative that eliminates harmful emissions, reduces fire hazards, and alleviates the burden of fuel collection. By harnessing the sun's energy, solar cookers provide a cost-effective and eco-friendly way to prepare meals without exposing individuals to toxic fumes and physical dangers.

While challenges such as weather dependency and cultural acceptance exist, continuous innovations in solar cooker design, thermal storage, and hybrid cooking technologies are making solar cooking more efficient and accessible. Governments, NGOs, and international organizations are increasingly promoting solar cooking initiatives, integrating them into public health and environmental programs to combat the health risks of traditional cooking. (Rosa G et al., 2014)

4. HOW SOLAR COOKING REDUCES INDOOR AIR POLLUTION

a. Reduction of Household Emissions and Particulate Matter

Indoor air pollution is a severe issue in households that rely on biomass fuels for cooking. Traditional stoves and open fires release pollutants such as carbon monoxide (CO), nitrogen oxides (NO_2), sulfur dioxide (SO_2), and fine particulate matter (PM2.5), all of which pose serious health risks. The transition to solar cooking provides a direct and effective solution by eliminating fuel combustion, thus significantly reducing household emissions and improving indoor air quality.

A major benefit of solar cooking is its zero-emission process. Unlike wood-burning stoves, which emit carbon dioxide, methane, and black carbon, solar cookers rely entirely on sunlight, which is a renewable and clean energy source. This not only improves health outcomes but also contributes to global efforts to reduce greenhouse gas emissions and combat climate change. (Albalak R et al., 2001)

The impact of indoor air pollution is particularly severe in rural communities, where ventilation is often poor, leading to prolonged exposure to toxic fumes. Women and children, who spend hours inside poorly ventilated kitchens, suffer the most from these emissions. Studies have shown that households using firewood for cooking experience indoor pollution levels 10–20 times higher than the recommended limits set by the World Health Organization (WHO).

Beyond individual health benefits, reducing household emissions has broader environmental advantages. Black carbon, a major byproduct of burning biomass, is a powerful contributor to global warming and has been linked to the melting of glaciers, particularly in regions like the Himalayas. By replacing biomass stoves with solar cookers, families can directly contribute to lowering atmospheric pollution, improving both public health and environmental sustainability.

b. Comparative Analysis of Indoor Air Quality in Solar vs. Traditional Cooking Households

Numerous studies have compared indoor air quality in households using traditional stoves versus those using solar cookers. The findings overwhelmingly support the conclusion that solar cooking significantly improves indoor air conditions, reducing the harmful pollutants that lead to respiratory and cardiovascular diseases. (Sudha Singh et al., 2014)

i. Carbon Monoxide Levels

Carbon monoxide is a deadly gas that is colorless, odorless, and highly toxic. In traditional cooking households, CO levels often exceed WHO's recommended safety limit of 9 ppm (parts per million). Studies in Latin America have found that women cooking with wood-burning stoves are exposed to CO levels as high as 50 ppm, putting them at severe risk of poisoning.

In contrast, homes using solar cookers report virtually zero CO emissions, eliminating the risk of poisoning and long-term respiratory damage. This significant reduction has been documented in field studies across Sub-Saharan Africa, where families transitioning to solar cooking have seen marked improvements in overall health.

ii. Particulate Matter (PM2.5) Reduction

Fine particulate matter, or PM2.5, is a primary contributor to respiratory diseases. These tiny particles can penetrate deep into the lungs and even enter the bloodstream, leading to asthma, lung infections, and cardiovascular conditions. Traditional biomass stoves generate high levels of PM2.5, often exceeding 500 $\mu g/m^3$ (micrograms per cubic meter)—a level that is considered highly hazardous to human health.

Comparative studies have shown that households using solar cookers experience a 75–90% decrease in PM2.5 exposure. A large-scale study in Nepal found that replacing firewood with solar cooking led to a sharp decline in respiratory illnesses among women and children. Similarly, research in Ethiopia demonstrated that households using solar cookers experienced fewer cases of pneumonia and lung disease, reducing hospital visits and healthcare costs.

iii. Health Benefits of Cleaner Indoor Air

Beyond measurable reductions in pollutants, families who transition to solar cooking report fewer instances of chronic coughing, eye irritation, and breathing difficulties. In Guatemala, a study comparing health outcomes in solar-cooking villages found that cases of bronchitis dropped by 40%, while symptoms of eye irritation reduced by 50%.

Moreover, the long-term impact of cleaner indoor air is particularly beneficial for children. Exposure to household smoke at an early age has been linked to stunted lung development, reduced cognitive function, and increased susceptibility to respiratory infections. By switching to solar cooking, parents can significantly improve their children's health and development, ensuring a better quality of life.

The overwhelming evidence from comparative studies underscores the urgent need to promote solar cooking as a scalable solution to indoor air pollution. Governments and NGOs must focus on educational campaigns, financial incentives, and community programs to increase adoption, particularly in regions where indoor air pollution remains a critical public health issue.

Table 2. Comparison between Traditional and Solar cookers

Aspect	Traditional Stoves	Solar Cookers
CO levels	Up to 50 ppm (unsafe)	Almost zero
PM2.5	>500 μg/m³(hazardous)	75-90% lower
Respiratory health	High illness risk	Improved health
Other symptoms	Coughing, eye irritation	Reduced symptoms
Study outcomes	High disease rates	40% less bronchitis, 50 less eye irritation

5. HEALTH BENEFITS OF SOLAR COOKING

Solar cooking is an innovative and sustainable method of food preparation that harnesses the power of the sun to cook meals. Unlike traditional cooking methods that rely on biomass, coal, or gas, solar cooking eliminates harmful emissions, reduces fuel dependency, and minimizes health risks. It has significant health benefits, particularly in regions where people suffer from the adverse effects of indoor air pollution, exposure to toxic fumes, and unsafe cooking environments. This section explores the key health advantages of solar cooking, including reduced respiratory illnesses, improved nutritional retention, and lower risks of burns and accidents. (Padonou E.A et al., 2022)

a. Reduction in Respiratory and Cardiovascular Diseases

One of the most significant health benefits of solar cooking is the reduction in respiratory and cardiovascular diseases caused by traditional cooking fuels. When biomass fuels such as firewood, charcoal, and animal dung are burned, they release harmful pollutants, including fine particulate matter (PM2.5), carbon monoxide, nitrogen oxides, and volatile organic compounds. These pollutants accumulate indoors, leading to household air pollution, which has been linked to chronic respiratory diseases such as asthma, bronchitis, and chronic obstructive pulmonary disease (COPD).

Solar cookers, on the other hand, produce no smoke or toxic emissions, ensuring that families, particularly women and children who spend long hours cooking, are not exposed to hazardous pollutants. Studies have shown that prolonged exposure to indoor air pollution significantly increases the risk of lung infections, tuberculosis, and lung cancer. By eliminating smoke and toxic fumes, solar cooking contributes to healthier lungs, improved breathing, and a lower risk of cardiovascular diseases, including high blood pressure, stroke, and heart attacks.

b. Prevention of Carbon Monoxide Poisoning

Traditional cooking methods, especially those involving wood stoves, coal, and kerosene, release carbon monoxide (CO), a highly toxic and odorless gas that can be deadly when inhaled in high concentrations. Carbon monoxide poisoning leads to symptoms such as dizziness, headaches, nausea, confusion, and, in severe cases, unconsciousness or death. Pregnant women exposed to carbon monoxide are at a higher risk of complications such as low birth weight and developmental disorders in infants.

Since solar cookers do not burn fuel, they eliminate the risk of carbon monoxide poisoning entirely. This is particularly beneficial in poorly ventilated homes where toxic gases from cooking fuels can accumulate rapidly. Using solar cooking systems ensures that individuals are not exposed to life-threatening carbon monoxide emissions, thereby promoting safer and healthier households.

c. Retention of Nutritional Value in Food

Another crucial health benefit of solar cooking is the preservation of nutrients in food. Traditional cooking methods often involve high temperatures and prolonged cooking times, which can lead to the loss of essential vitamins, minerals, and antioxidants. Frying, boiling, and grilling at high heat can degrade nutrients such as vitamin C, folate, and certain proteins, reducing the overall nutritional value of meals. (Koh et al., 2016)

Solar cookers, however, use gentler and slower cooking processes, which help retain more nutrients in food. Low-temperature cooking preserves water-soluble vitamins and prevents the breakdown of essential nutrients. This is particularly important in regions where malnutrition is a concern, as it ensures that food retains its full nutritional potential, contributing to better overall health. (Bhave & Thakare, 2018)

d. Healthier Cooking Practices with Less Oil Usage

Traditional frying methods require a significant amount of cooking oil, which can lead to the formation of harmful trans fats and carcinogenic compounds when heated at high temperatures. Frequent consumption of fried foods has been linked to obesity, heart disease, and increased cholesterol levels.

Solar cooking encourages oil-free or low-oil cooking, as the slow and even heating process allows food to cook in its natural moisture without the need for excessive fats. This helps reduce the intake of unhealthy oils and lowers the risk of diet-related diseases such as diabetes, hypertension, and cardiovascular disorders. Additionally, solar cooking enhances the natural flavors of food, reducing the need for excessive seasoning or artificial additives.

e. Reduction in Waterborne Diseases

Access to clean drinking water is a major challenge in many developing regions, and contaminated water is a leading cause of gastrointestinal infections, diarrhea, and other waterborne diseases. Boiling water is one of the most effective ways to eliminate harmful pathogens, but in fuel-scarce areas, families may struggle to purify water regularly due to the high cost of firewood or gas.

Solar cookers provide a cost-effective and sustainable solution for water purification. By heating water to boiling temperatures using solar energy, they effectively kill bacteria, viruses, and parasites, making water safe for consumption. This simple yet impactful benefit can help reduce cases of diarrhea, typhoid, cholera, and other waterborne illnesses, particularly in communities with limited access to clean water sources.

f. Improved Maternal and Child Health

The health benefits of solar cooking extend to maternal and child health, as women and young children are the most affected by the negative consequences of traditional cooking methods. Pregnant women exposed to indoor air pollution face a higher risk of preterm births, low birth weight babies, and complications during pregnancy. Additionally, malnourished mothers who consume nutrient-deficient diets may experience anemia and other health issues.

By promoting clean cooking and improved nutrition, solar cooking contributes to better maternal health outcomes. Children, especially in their early developmental years, benefit from improved air quality and higher nutrient retention in food, leading to stronger immune systems, better cognitive development, and overall healthier growth.

6. ENVIRONMENTAL AND SUSTAINABILITY BENEFITS OF SOLAR COOKING

a. Impact of Solar Cooking on Deforestation and Carbon Footprint Reduction

Deforestation is one of the most pressing environmental issues worldwide, and one of its major drivers is the demand for firewood and charcoal used in traditional cooking methods. In many developing regions, wood remains the primary cooking fuel, leading to massive tree loss, habitat destruction, and soil degradation. Solar cooking, which entirely eliminates the need for firewood, presents an effective solution to reduce deforestation and its associated environmental impacts.

b. Deforestation and Its Consequences

According to studies by the Food and Agriculture Organization (FAO), nearly 2.4 billion people worldwide rely on biomass fuels, such as firewood, charcoal, and crop residues, for cooking. The high demand for firewood leads to unsustainable logging, depleting forest resources at an alarming rate. In Africa, deforestation caused by wood fuel collection accounts for nearly 40% of all tree loss, while in regions like the Amazon, charcoal production has contributed significantly to forest degradation.

Solar cooking provides a sustainable alternative that directly addresses this crisis. By eliminating reliance on firewood and charcoal, it allows forests to regenerate, preserving critical ecosystems and preventing further environmental degradation. A case study in Kenya revealed that families who adopted solar cooking reduced their firewood consumption by 70%, leading to noticeable improvements in local reforestation efforts.

c. Reduction in Carbon Emissions

Traditional biomass stoves are major contributors to carbon emissions, which drive global climate change. The combustions of firewood releases carbon dioxide (CO_2), methane (CH_3), and black carbon, all of which give to global warming. Black carbon, in particular, is a short-lived but highly potent climate pollutant, with a global warming potential 680 times greater than CO_2 over a 20-year period.Solar cooking completely eliminates these emissions. Studies indicate that if just one million households switched from wood-burning stoves to solar cookers, it would prevent the release of approximately 6 million metric tons of CO_2annually—equivalent to removing 1.5 million cars from the road.

The environmental benefits of solar cooking extend beyond individual households. In large-scale community settings, such as schools and refugee camps, the adoption of solar cooking has led to substantial reductions in carbon emissions. For instance, the Kakuma Refugee Camp in Kenya introduced solar cookers to thousands of families, significantly cutting down firewood consumption and reducing the camp's overall carbon footprint.

d. Role of Solar Cooking in Achieving Sustainable Development Goals (SDGs)

The United Nations' Sustainable Development Goals (SDGs) aim to create a more sustainable and equitable world by 2030. Solar cooking aligns with several of these goals, particularly those related to energy, health, climate action, and gender equality.

- **SDG 7: Affordable and Clean Energy**

Solar cooking directly supports SDG 7, which focuses on ensuring universal access to affordable, reliable, and sustainable energy. Unlike traditional fuels, solar energy is free, abundant, and renewable, making it an ideal solution for off-grid communities. By harnessing the sun's power, solar cookers provide a cost-effective alternative that reduces dependency on expensive and polluting fuels like kerosene and charcoal.

- **SDG 3: Good Health and Well-Being**

As discussed earlier, traditional cooking methods contribute to severe health problems, including respiratory diseases, eye infections, and burns. Solar cooking eliminates these risks, significantly improving overall health and well-being. By reducing exposure to harmful smoke and toxic emissions, it directly supports SDG 3, which aims to ensure healthy lives for all.

- **SDG 13: Climate Action**

Solar cooking plays a crucial role in climate change mitigation, aligning with SDG 13. By reducing deforestation and cutting down carbon emissions, it helps slow the pace of global warming. In regions particularly vulnerable to climate change—such as coastal communities facing rising sea levels or arid regions suffering from desertification—solar cooking provides a sustainable solution that lessens human impact on the environment.

Additionally, international organizations have recognized the potential of solar cooking in carbon credit programs. By quantifying the emissions reductions achieved through solar cooking adoption, carbon credits can be sold to industries seeking to offset their emissions. This model not only benefits the environment but also provides financial incentives for communities to adopt clean cooking technologies.

- **SDG 5: Gender Equality**

Women in many developing countries bear the burden of cooking and fuel collection. On average, women in rural areas spend 15–20 hours per week gathering firewood, time that could be used for education, employment, or entrepreneurship. By adopting solar cooking, households can eliminate this labor-intensive task, empowering women to engage in more productive activities.

A study in Nepal found that women in solar-cooking households had more time to pursue income-generating activities, leading to improved economic stability for their families. In addition, reducing exposure to smoke and burns enhances women's health and quality of life, further promoting gender equality. (Achudume, 2009).

e. Scaling Up Solar Cooking for Maximum Impact

To fully leverage solar cooking's potential in achieving the Sustainable Development Goals, governments and international agencies must scale up investments in solar cooking technologies.

This includes:

- Subsidies and financial incentives for low-income households
- Public awareness campaigns highlighting the benefits of solar cooking
- Incorporating solar cooking into national energy policies
- Research and development to improve efficiency and affordability

7. SAFETY FEATURES IN THE DESIGN OF SOLAR COOKERS

a. Heat-Resistant Materials and Insulated Cooking Chambers

One of the most critical aspects of solar cooker design is ensuring safety by using heat-resistant materials and insulated cooking chambers. Unlike traditional cooking methods that involve open flames, which pose risks of burns, fire hazards, and exposure to toxic fumes, solar cookers eliminate these dangers by incorporating materials specifically designed to withstand high temperatures while maintaining structural integrity. Modern solar cookers utilize borosilicate glass, stainless steel, and heat-resistant plastics to enhance both durability and safety. Borosilicate glass, often used in the cooking chamber or as a transparent lid, is highly resistant to thermal shock, preventing cracks and breakage even when exposed to extreme heat fluctuations. This reduces the risk of injuries caused by shattered glass. Similarly, stainless steel is used for reflectors and inner cooking surfaces because of its ability to withstand prolonged exposure to sunlight without degrading. Heat-resistant plastics and silicone components are strategically placed on handles, outer casings, and locking mechanisms to prevent users from accidentally touching hot surfaces, ensuring safe handling.

Insulated cooking chambers play a crucial role in both energy efficiency and user safety. Unlike conventional cooking appliances that generate excessive external heat, solar cookers are designed with double-walled insulation that helps trap heat within the cooking chamber while keeping the outer surfaces cool. This not only enhances cooking efficiency by maintaining stable temperatures but also reduces the risk of accidental burns, making solar cookers especially safe for families with children. In regions where people are accustomed to cooking over open fires, transitioning to a solar cooker provides a safer alternative by eliminating direct exposure to flames. Furthermore, insulation materials such as vacuum-sealed panels, fiberglass, and mineral wool are integrated into some high-end solar cookers, preventing heat loss even in fluctuating weather conditions. This allows food to cook evenly and reduces the need for constant monitoring, unlike traditional cooking, where fluctuations in fire intensity can cause inconsistent results. Some advanced models even incorporate phase change materials (PCMs) that store excess heat during peak sunlight hours and release it gradually, keeping food warm for extended periods after sunset.

b. Sunlight Concentration and Reflector Safety Mechanisms

The effectiveness of a solar cooker largely depends on its ability to concentrate sunlight onto the cooking chamber. However, this feature also presents certain safety challenges, including excessive glare, accidental burns, and unintended heat build-up. To mitigate these risks, modern solar cookers incorporate innovative reflector designs and safety mechanisms that enhance usability without compromising safety. Reflectors are typically made from polished stainless steel or anodized aluminum, materials that efficiently direct sunlight while resisting heat damage and corrosion. Unlike mirrors, which can shatter easily and create hazardous shards, stainless steel and aluminum offer a more durable and safer alternative. Despite their efficiency, poorly positioned reflectors can cause intense glare, which may lead to eye strain or long-term vision damage. To address this, many solar cookers are designed with angled reflectors that direct sunlight precisely onto the cooking surface without dispersing excess light outward. Some high-end models even feature anti-glare coatings on the reflectors, which reduce brightness intensity while maintaining optimal heat absorption. (BishwashPaneru, 2024)

In addition to glare reduction, burn prevention is a key focus in the design of solar cookers. Since solar cookers can reach temperatures exceeding 200–300°C (392–572°F), users must be protected from accidental contact with heated components. To enhance safety, some models incorporate protective heat shields around the reflectors, creating a buffer zone that prevents users from directly touching hot surfaces. These shields not only reduce the risk of burns but also help contain heat within the cooking chamber, improving efficiency. Many modern solar cookers are also equipped with cool-touch handles made of silicone or other heat-resistant materials, allowing users to open and close the cooker safely without gloves. Furthermore, some solar cookers include locking mechanisms for the reflectors, preventing unintended shifts in position that could cause excessive heat concentration in one area. This is particularly important in parabolic solar cookers, which generate high temperatures quickly and require precise reflector positioning for safe operation.

Some hybrid solar cookers come equipped with temperature sensors that monitor internal heat levels and trigger cooling mechanisms if necessary. Additionally, automatic sun-tracking systems have been developed to ensure that reflectors remain optimally positioned throughout the day without manual adjustments. These systems improve efficiency while also reducing the risk of accidental overexposure to heat.

8. SAFETY PRECAUTIONS AND GUIDELINES FOR USING SOLAR COOKERS

a. Best Practices to Prevent Burns and Glare Injuries

While solar cookers eliminate the dangers of open flames and fuel combustion, users must still follow certain safety precautions to avoid burns and glare-related injuries and irritation caused by exposure to glass fibers used in certain cooker designs. One of the most important practices is handling cooking pots and food containers carefully, as they can become extremely hot during the cooking process. Unlike conventional stoves that cool down quickly after being turned off, solar cookers retain heat for an extended period due to their insulated design. This means that even after the sun sets, the cooking chamber and its contents may remain hot for hours. To prevent burns, users should always use protective gloves or thick cloth when handling pots, lids, or trays inside the cooker. Some modern solar cookers come with built-in insulated handles, but in many traditional models, these features are not present, requiring extra caution from users.

Materials must be chosen carefully. In the past, asbestos sheets where sometime used as insulation because of their heat resistance but these causes serious health conditions such as asbestosis, lung cancer and mesothelioma. Therefore the use of asbestos in solar cooker is strictly prohibited and safer alternative such as fiber glass, mineral wool or ceramic insulations are used. Another significant risk when using solar cookers is the potential for eye strain or long-term vision damage caused by intense glare from reflective surfaces. Most solar cookers rely on highly polished aluminum or stainless-steel panels to concentrate sunlight onto the cooking chamber. When positioned incorrectly or used without proper protection, these panels can reflect excessive sunlight directly into the eyes, leading to discomfort, temporary blindness, or, in extreme cases, retinal damage. To minimize this risk, users should avoid looking directly into the concentrated light reflections from parabolic or panel cookers. Wearing UV-protective sunglasses can significantly reduce eye strain, especially in areas with high sunlight intensity. Additionally, positioning the solar cooker so that it faces away from direct eye contact with bystanders or passersby can help reduce accidental glare exposure.

The placement of the solar cooker itself is another crucial safety consideration. Solar cookers should always be positioned on a stable, flat surface to prevent accidental tipping or shifting. Uneven surfaces can cause the cooker to tilt unexpectedly, leading to spills, burns, or even damage to the cooker itself. In areas with high wind conditions, solar cookers with large reflectors may be susceptible to being blown over. To prevent this, users should secure the cooker with weights or stakes, especially when using it outdoors in open spaces. Some solar cookers come with built-in stabilizing bases, but for those that do not, extra precautions must be taken to keep the cooker anchored.

Children and pets should be kept at a safe distance from solar cookers while they are in use. Although solar cookers are generally safer than traditional stoves, curious children may accidentally touch hot surfaces or knock over the cooker, leading to injuries. Educating family members, especially young children, about the importance of not touching or playing around the solar cooker is an essential safety measure. In community settings where multiple families share solar cookers, clear safety guidelines should be established to ensure that everyone understands how to use the equipment responsibly.

For those using parabolic solar cookers, which reach extremely high temperatures quickly, extra care must be taken to avoid direct skin exposure to concentrated sunlight. The focal point of a parabolic cooker can reach over 300°C (572°F), posing a significant burn risk if hands or other objects come into contact with it. Users should always approach the cooking pot from the side rather than reaching over the cooker to avoid accidental burns. Additionally, keeping a lid on the pot while cooking helps contain heat and reduces the risk of accidental spills. Some advanced solar cookers include automated tracking systems that adjust the position of the reflectors throughout the day, eliminating the need for manual adjustments and reducing the chances of accidental exposure to intense sunlight. By following these best practices, users can ensure a safe and injury-free cooking experience while enjoying the benefits of solar energy.

b. Food Hygiene and Temperature Management in Solar Cooking

Maintaining proper food hygiene and temperature management is essential when using solar cookers to ensure that meals are safe for consumption. Unlike conventional stoves, which provide consistent and controllable heat, solar cookers rely on varying levels of sunlight, which can sometimes lead to fluctuating cooking temperatures. If food is not heated adequately, harmful bacteria and pathogens may survive, increasing the risk of foodborne illnesses. To prevent this, users must ensure that food reaches and maintains safe internal temperatures before consumption. According to food safety guidelines, poultry and meat should reach a minimum internal temperature of 75°C (165°F), while soups and stews should be heated to at least 70°C (158°F) to eliminate bacteria.

One way to ensure that food is cooked thoroughly in a solar cooker is by using a food thermometer. A simple digital or analog thermometer can help users verify that their meals have reached the required temperature before serving. In addition to monitoring cooking temperatures, users should also be mindful of the cooking duration. Since solar cooking relies on gradual heat accumulation, food should be given ample time to cook completely. Dishes that require prolonged simmering, such as beans or rice, may take longer than they would on a conventional stove. To optimize cooking times, it is recommended to preheat the solar cooker by exposing it to direct sunlight for 15–30 minutes before placing food inside. This helps the cooker reach an ideal starting temperature and reduces overall cooking time.

Another crucial aspect of food hygiene is preventing contamination during the cooking process. Since solar cookers do not involve direct flames or high-pressure cooking, users must take extra precautions to keep cooking utensils and food containers clean. Before placing food in the cooker, all pots, trays, and lids should be thoroughly washed to remove any bacteria or residues. Additionally, raw meat should be kept separate from other ingredients to prevent cross-contamination. In community settings where multiple people use the same solar cookers, implementing designated cooking stations for different types of food can help maintain hygiene standards.

Storage of cooked food is another important consideration. Since solar cookers do not have built-in cooling mechanisms, food should not be left inside the cooker for extended periods after cooking is complete. In warm climates, leaving cooked food in the solar cooker for too long can create an environment for bacterial growth. To prevent this, cooked meals should be consumed immediately or stored in a covered container in a cool place. If leftovers are to be saved for later, they should be refrigerated or reheated to a safe temperature before consumption.

In some cases, solar cooking can also be used for pasteurization, an important process for ensuring safe drinking water and milk consumption. Water can be effectively pasteurized in a solar cooker by heating it to 65°C (149°F) for at least six minutes, which kills most harmful bacteria and parasites. This is particularly beneficial in regions where access to clean drinking water is limited. Some solar cookers are designed with built-in indicators that change color when water reaches pasteurization temperature, helping users ensure safe water consumption. Similarly, milk and fruit juices can be pasteurized in solar cookers to prevent bacterial contamination and extend shelf life.

9. INNOVATIONS IN SOLAR COOKER DESIGN TO ENHANCE SAFETY AND EFFICIENCY

a. Hybrid Solar Cookers with Thermal Energy Storage

One of the most significant advancements in solar cooker design is the development of hybrid models that incorporate thermal energy storage. Traditional solar cookers rely entirely on direct sunlight, meaning that cooking can only take place when the sun is shining. This dependence on daylight limits their usability, particularly in cloudy conditions, early mornings, evenings, or during rainy seasons. To overcome this limitation, researchers and engineers have introduced hybrid solar cookers that store thermal energy for later use. These systems utilize various heat-retention materials, such as phase change materials (PCMs), molten salts, and heat-retentive bricks, to capture and store excess solar energy during peak sunlight hours. This stored heat is then gradually released, allowing users to continue cooking after sunset or when sunlight is insufficient.

Thermal energy storage materials work by absorbing heat during the day and maintaining high temperatures over an extended period. For example, PCMs like paraffin wax or salt hydrates can store heat at a consistent temperature and release it slowly when needed, ensuring a stable cooking environment. This technology is particularly useful for slow-cooked meals that require consistent heat over several hours, such as stews, beans, or baked goods. Some advanced hybrid solar cookers integrate insulated storage compartments, allowing users to cook food during the day and keep it warm for consumption later in the evening. This innovation eliminates one of the biggest drawbacks of conventional solar cookers limited cooking times and significantly enhances their practicality.

In addition to improving convenience, hybrid solar cookers also increase safety. Since they do not require an open flame or continuous fuel supply, they eliminate fire hazards associated with traditional cooking. The ability to store heat for later use also reduces the need for users to frequently adjust or reposition the cooker, which can sometimes lead to accidental burns or injuries. Moreover, thermal energy storage allows families to prepare meals during the most convenient parts of the day rather than being restricted to cooking only when sunlight is at its peak. This flexibility is especially beneficial for households where members are engaged in work or school during daylight hours and need to prepare food in the evening.

Several real-world applications have demonstrated the success of hybrid solar cookers. In India, large-scale solar kitchens have incorporated thermal energy storage to provide continuous meal service for schools and religious institutions. Similarly, in Africa, community based solar cooking initiatives have leveraged thermal storage technology to ensure food security in rural areas with inconsistent sunlight exposure. Ongoing research is focused on making these systems more affordable and accessible by using cost-effective storage materials and optimizing insulation techniques. As these innovations continue to evolve, hybrid solar cookers are expected to become a mainstream alternative to traditional cooking methods, offering a sustainable, safe, and efficient solution for households worldwide. (Prashant Saini et al., 2023)

b. Automation and Smart Tracking Systems in Solar Cooking

Another groundbreaking innovation in solar cooking is the integration of automation and smart tracking systems. Traditional solar cookers require manual adjustments to ensure they remain properly aligned with the sun. This process can be time-consuming, especially for users who need to frequently reposition their cookers throughout the day to maximize heat absorption. To address this challenge, engineers have developed automatic solar tracking systems that adjust the position of the cooker in real-time, ensuring optimal sunlight exposure without the need for user intervention.

Solar tracking technology works by using light sensors, motors, and control algorithms to continuously follow the sun's movement across the sky. These systems can be either single-axis or dual-axis. Single-axis trackers adjust the cooker's angle from east to west, while dual-axis trackers fine-tune the position both horizontally and vertically for maximum solar efficiency. This automation significantly increases cooking efficiency by maintaining a constant focus of sunlight on the cooking chamber, resulting in faster cooking times and higher temperatures. Some advanced models even use GPS-based tracking to pre-program optimal positioning based on geographical location and seasonal variations in sunlight.

Beyond improving efficiency, smart tracking systems also enhance safety. Manually repositioning a solar cooker can sometimes expose users to intense sunlight reflections, increasing the risk of glare-related eye injuries. With automated tracking, users can place food inside the cooker and let the system handle the adjustments, reducing direct interaction with potentially hazardous reflective surfaces. Additionally, automation minimizes the need for frequent handling of hot cookware, decreasing the risk of burns.

Smart solar cookers are also being equipped with digital temperature sensors and wireless monitoring features. These innovations allow users to track cooking progress remotely through smartphone applications or digital displays. Temperature sensors ensure that food is cooked at safe temperatures by providing real-time data and alerts if heat levels drop too low. Some models even include automatic shut-off mechanisms that prevent overheating or burning of food, adding an extra layer of safety.

The implementation of Internet of Things (IoT) technology in solar cooking is another promising development. Future smart solar cookers may come with integrated AI-powered cooking assistants that can suggest optimal cooking times, adjust heat settings based on the type of food being prepared, and provide alerts when the meal is ready. This level of automation could make solar cooking even more user-friendly, attracting a broader audience, including urban households and tech-savvy consumers. (Selvamurugan C et al., 2025)

Several pilot projects have already demonstrated the potential of automated solar cookers. In China, researchers have developed smart solar cookers with built-in microcontrollers that autonomously adjust cooking angles. In Europe, experimental models are being tested with Bluetooth connectivity, allowing users to control their cookers remotely. These innovations are making solar cooking more attractive to modern consumers by reducing manual effort, improving cooking precision, and ensuring safety.

Looking ahead, further improvements in battery storage, AI integration, and lightweight materials will continue to refine the efficiency and convenience of solar cookers. As these innovations become more mainstream, solar cooking will no longer be seen as just an alternative but as a primary, reliable, and safe cooking method for both rural and urban communities. By combining thermal storage, automation, and smart tracking systems, the next generation of solar cookers is set to revolutionize the way people prepare food while reducing environmental impact and enhancing user safety. (Selvakumar P et al., 2024)

10. CHALLENGES AND LIMITATIONS OF SOLAR COOKING

Solar cooking is a sustainable and eco-friendly alternative to traditional cooking methods, offering benefits such as reduced air pollution and decreased reliance on non-renewable fuels. However, despite these advantages, several challenges and limitations hinder its widespread adoption. Dependence on weather conditions, long cooking times, limited versatility, high initial costs, cultural resistance, and maintenance difficulties all contribute to the slow acceptance of solar cooking technology. Addressing these barriers is essential for improving solar cooking systems and ensuring their viability for diverse communities.

a. Dependence on Sunlight and Weather Conditions

One of the most significant challenges of solar cooking is its reliance on sunlight. Solar cookers perform best in areas with abundant sunshine, but their efficiency drops significantly in cloudy, rainy, or winter conditions. In regions with unpredictable weather, cooking times may be extended, and food may not reach the necessary temperatures for proper cooking. (Cuce & Cuce, 2013)

Additionally, solar cookers cannot function at night, meaning users must prepare meals during the day. This limitation is inconvenient for individuals who need to cook early in the morning or late at night. While some advanced solar cookers incorporate thermal storage systems to retain heat, these models are often expensive and not widely available. Hybrid solar cookers, which combine solar energy with alternative fuel sources, provide a potential solution, but they also add to the cost and complexity of the technology.

b. Longer Cooking Times and Inconvenience

Compared to conventional cooking methods, solar cookers require significantly longer cooking times. The heating process depends on sunlight intensity, meaning food may take hours to cook, especially in moderate sunlight conditions. Unlike gas or electric stoves, which provide instant heat, solar cookers require preheating and careful planning to ensure meals are ready on time.

This slow cooking process can be inconvenient, particularly for people with busy schedules who need quick meal preparation. It also poses challenges for large families or communities that need to cook multiple meals throughout the day. In emergency situations or disaster-prone regions, where fast meal preparation is necessary, the time-consuming nature of solar cooking can be a drawback.

c. Limited Cooking Capacity and Versatility

Solar cookers are primarily designed for boiling, steaming, and baking, but they are not well-suited for high-temperature cooking methods such as frying or deep-frying. Traditional cooking methods offer greater flexibility in preparing a variety of dishes, whereas solar cookers may limit meal options in certain regions.

Additionally, the cooking capacity of many solar cookers is relatively small. Box and panel solar cookers, which are commonly used in households, may not be large enough to cook meals for big families or community gatherings. This limitation makes them less practical for large-scale cooking needs, requiring multiple units or alternative cooking solutions.

d. High Initial Costs and Affordability Issues

Although solar cooking eliminates the need for fuel in the long run, the upfront cost of purchasing a solar cooker can be a significant barrier for low-income households. High-quality solar cookers that offer better heat retention and durability tend to be more expensive, making them inaccessible to economically disadvantaged communities.

Furthermore, the lack of subsidies, financial support, and incentives in many regions limits adoption. Many families prioritize immediate affordability over long-term benefits, opting for traditional cooking methods that require cheaper upfront investments. Expanding financial assistance programs, promoting local production of cost-effective solar cookers, and implementing government incentives could help overcome this economic barrier.

e. Cultural and Social Acceptance

Cultural norms and traditional cooking habits play a major role in the acceptance of solar cooking. In many societies, cooking with fire is deeply ingrained in daily life, and transitioning to a new method requires behavioral changes. Some communities may hesitate to adopt solar cooking due to unfamiliarity, skepticism about its effectiveness, or concerns about differences in taste and food texture.

Additionally, in certain regions, cooking is a social activity, and using a solar cooker may disrupt traditional cooking practices. Women, who are often responsible for meal preparation, may be reluctant to adopt solar cooking if they are not properly educated on its benefits. Awareness programs, community engagement initiatives, and demonstration projects are necessary to address misconceptions and encourage widespread adoption.

f. Maintenance and Durability Challenges

Solar cookers, especially those made from low-cost materials, may suffer from durability and maintenance issues. Exposure to harsh weather conditions, such as heavy rain, strong winds, and high humidity, can degrade certain types of solar cookers over time.

Some models require frequent maintenance, such as cleaning reflectors, replacing damaged insulation, or repairing fragile components. Users who lack technical knowledge or access to spare parts may find it difficult to maintain their solar cookers, leading to low adoption rates in some areas. To address this challenge, manufacturers must focus on designing durable, low-maintenance solar cookers that can withstand various environmental conditions.

g. Storage and Nighttime Cooking Limitations

Unlike gas or electric stoves, solar cookers do not provide continuous cooking options after sunset. This poses a challenge for individuals who need to prepare meals at night or during early morning hours. Some advanced solar cookers incorporate thermal storage systems, allowing them to retain heat for extended periods, but these models are often expensive and not widely available.

A potential solution is the development of hybrid solar cookers that use solar energy during the day and alternative energy sources at night. However, widespread implementation of such systems requires further research, funding, and infrastructure development. (Balachandran & Swaminathan, 2022)

CONCLUSION

Solar cooking presents a transformative approach to food preparation that addresses critical health and safety concerns associated with traditional cooking methods. By eliminating indoor air pollution, reducing reliance on hazardous fuels, and preventing fire-related accidents, solar cookers offer a cleaner and safer alternative, particularly in regions where biomass and fossil fuels are the primary energy sources. The health benefits of solar cooking are substantial, ranging from lower respiratory and cardiovascular risks to improved nutritional retention in food.

Beyond health, solar cooking contributes to environmental sustainability by reducing deforestation, lowering carbon emissions, and minimizing energy consumption. Advancements in solar cooker design continue to enhance efficiency, safety, and accessibility, making this technology a viable solution for households worldwide. However, challenges such as weather dependency, affordability, and cultural acceptance must be addressed through innovation, education, and policy support.

As the world transitions toward sustainable energy solutions, solar cooking stands out as a promising technology that promotes health, environmental conservation, and energy security. Widespread adoption, supported by research and investment, can significantly improve public health outcomes while mitigating the adverse effects of traditional cooking practices. Embracing solar cooking is not just an alternativeit is a necessary step toward a healthier and more sustainable future.

Although the existing evidence indicates the substantial health and environmental advantages of solar cooking, more research will be necessary to design cookers that are optimized for various climatic conditions, to improve thermal storage systems for night cooking or low-sunlight conditions, and to measure long-term health benefits in various populations. Research into cost reduction, training users, and cultural compatibility must also be conducted to facilitate widespread acceptance. Combination of solar cooking with hybrid renewable energy systems and enabling policy frameworks is one area that promises to be explored in the future.

REFERENCES

Achudume, A. C. (2009). Environmental health, development and economic empowerment of rural women in Nigeria. *Environment, Development and Sustainability, 11*, 459–469. DOI: 10.1007/s10668-007-9124-1

Albalak R, Bruce N, McCracken JP, Smith KR, De Gallardo T (2001). Indoor respirable particulate matter concentrations from an open fire, improved cookstove, and LPG/open fire combination in a rural Guatemalan community. Environ Sci Technol. Jul 1;35(13):2650-5. . PMID: 11452588.DOI: 10.1021/es001940m

Balachandran, S., & Swaminathan, J. (2022). Advances in Indoor Cooking Using Solar Energy with Phase Change Material Storage Systems. *Energies*, *15*, 8775. DOI: 10.3390/en15228775

Bhave, A. G., & Thakare, K. A. (2018). Development of a solar thermal storage cum cooking device using salt hydrate. *Solar Energy*, *171*, 784–789. DOI: 10.1016/j.solener.2018.07.018

Biran, A., Abbot, J., & Mace, R. (2004). Families and Firewood: A Comparative Analysis of the Costs and Benefits of Children in Firewood Collection and Use in Two Rural Communities in Sub-Saharan Africa. *Human Ecology*, *32*, 1–25. DOI: 10.1023/B:HUEC.0000015210.89170.4e

BishwashPaneru. (2024). BiplovPaneru, Vikram Alexander, Silvia Nova, Nawraj Bhattarai, RamhariPoudyal, Khem Narayan Poudyal, Mohan B. Dangi, John J. Boland(2024). Solar energy for operating solar cookers as a clean cooking technology in South Asia: A review. *Solar Energy*, *283*, 113004. DOI: 10.1016/j.solener.2024.113004

Cuce, E., & Cuce, P. M. (2013). A comprehensive review on solar cookers. *Applied Energy*, *102*, 1399–1421. DOI: 10.1016/j.apenergy.2012.09.002

Farmer N, Cotter EW (2021). Well-Being and Cooking Behavior: Using the Positive Emotion, Engagement, Relationships, Meaning, and Accomplishment (PERMA) Model as a Theoretical Framework. Front Psychol. Apr 12;12:560578. doi: .DOI: 10.3389/fpsyg.2021.560578

Funk, P. A., & Larson, D. L. (1998). Parametric model of solar cooker performance. *Solar Energy*, *62*(1), 63–68. DOI: 10.1016/S0038-092X(97)00074-1

Kimambo, C. (2007). Development and performance testing of solar cookers. *Journal of Energy in Southern Africa*, *18*, 41–51. DOI: 10.17159/2413-3051/2007/v18i3a3384

Koh, K., Kim, Y.-J., & Kim, H.-S. (2016). Nutrition Analyses of Nutrition Bar Cooked with a Solar Box-cooker. *Korean Journal of Human Ecology.*, *25*, 487. DOI: 10.5934/kjhe.2016.25.4.487

Liu, W., Luo, X., Huang, Y., Zhao, M., Liu, T., Wang, J., & Feng, F. (2023). Influence of cooking techniques on food quality, digestibility, and health risks regarding lipid oxidation. *Food Research International*, *167*, 112685. DOI: 10.1016/j.foodres.2023.112685 PMID: 37087258

Mehta K, Thrikutam N, Hoyte-Williams PE, Falk H, Nakarmi K, Stewart B. (2023) Epidemiology and Outcomes of Cooking- and Cookstove-Related Burn Injuries: A World Health Organization Global Burn Registry Report. J Burn Care Res. May 2;44(3):508-516. doi: .DOI: 10.1093/jbcr/irab166

Ning, Ke; Zhou, Yan-Yan; Zhang, Ning; Sun, Xue-Jun; Liu, Wen-Wu; Han, Cui-Hong MD (2020). Neurocognitive sequelae after carbon monoxide poisoning and hyperbaric oxygen therapy. Medical Gas Research 10(1):p 30-36, Jan–Mar. | DOI: DOI: 10.4103/2045-9912.279981

Padonou, E. A., Akabassi, G. C., Akakpo, B. A., & Sinsin, B. (2022). Importance of solar cookers in women's daily lives: A review. *Energy for Sustainable Development*, *70*, 466–474. DOI: 10.1016/j.esd.2022.08.015

Panwar, N. L., Kaushik, S. C., & Kothari, S. (2012). State of the art of solar cooking: An overview. *Renewable & Sustainable Energy Reviews*, *16*, 3776–3785. DOI: 10.1016/j.rser.2012.03.026

Rodríguez-Ayala M, Sandoval-Insausti H, Bayán-Bravo A, Banegas JR, Donat-Vargas C, Ortolá R, Rodríguez-Artalejo F, Guallar-Castillón P.(2022) Cooking Methods and Their Relationship with Anthropometrics and Cardiovascular Risk Factors among Older Spanish Adults. Nutrients. Aug 20;14(16):3426. doi: .DOI: 10.3390/nu14163426

Rosa, G., Majorin, F., Boisson, S., Barstow, C., Johnson, M., Kirby, M., & Clasen, T. (2014). Assessing the impact of water filters and improved cook stoves on drinking water quality and household air pollution: A randomized controlled trial in Rwanda. *PLoS One*, *9*, e91011. Advance online publication. DOI: 10.1371/journal.pone.009101 PMID: 24614750

Saini, Prashant, Pandey, Sushant, Goswami, Shruti, & Dhar, Atul, M.E. Mohamed, Satvasheel Powar. (2023). Experimental and numerical investigation of a hybrid solar thermal-electric powered cooking oven. *Energy*, *280*, 128188. DOI: 10.1016/j.energy.2023.128188

Selvakumar, P., Muthusamy, S., Satishkumar, D., Vigneshkumar, P., Selvamurugan, C., & Kumar, P. S. (2024). *AI-Powered Tools, IGI Global, Mar.* IGI Global., DOI: 10.4018/979-8-3693-2615-2.ch002

Selvamurugan, C., Delna Renish, M., Ishwarya, F.J., & Hisha Sherin K.(2025). Social Dining Platform:Dining Platforms and their Impact on Food Culture and Community Building" IGI Global, Feb, DOI: .DOI: 10.4018/979-8-3693-9025-2.ch008

Singh, Sudha, Gupta, Gyan Prakash, & Kumar, Bablu, U.C. Kulshrestha. (2014). Comparative study of indoor air pollution using traditional and improved cooking stoves in rural households of Northern India. *Energy for Sustainable Development*, *19*, 1–6. DOI: 10.1016/j.esd.2014.01.007

Yu, G., Zhang, G., Poslad, S., Fan, Y., & Xu, X. (2024). A study of quantifying the influence of kitchen human activity on indoor air quality dynamics. *Environmental Pollution*, *362*, 124900. DOI: 10.1016/j.envpol.2024.124900 PMID: 39260554

Compilation of References

A. Simon Prabu, V. Chithambaram, M. Anto Bennet, S. Shanmugan, and B. Janarthanan, "Performance of solar cooker with evacuated tubes and photovoltaic panels with phase change materials", *Energy Efficiency*, vol. 16, n(o) 7, p. 72, August 2023, doi: .DOI: 10.1007/s12053-023-10151-3

Abed, F. (2011). Experimental investigation of thermal performance of solar cooker with reflector. *European Journal of Scientific Research*, *56*, 112–123.

Abhishek, S. A. X. E. N. A., & Mehmet, K. A. R. A. K. I. L. C. I. K. (2017). Performance evaluation of a solar cooker with low-cost heat storage material. *International Journal of Sustainable and Green Energy*, *6*(4), 57–63.

Abhishek, S. A. X. E. N. A., & Nitin, A. G. A. R. W. A. L. (2018). Performance characteristics of a new hybrid solar cooker with air duct. *Solar Energy*, *159*, 628–637.

Achudume, A. C. (2009). Environmental health, development and economic empowerment of rural women in Nigeria. *Environment, Development and Sustainability*, *11*, 459–469. DOI: 10.1007/s10668-007-9124-1

ADEME. Zeleke & HAMEER, Sameer. (2018). Design, construction and performance evaluation of a Box type solar cooker with a glazing wiper mechanism. AIMS energy vol. 6, no 1

Adewumi, S. D., Abdulrahim, A. T., Adeleke, A. A., Ikubanni, P. P., & Agboola, O. O. (2019). Performance evaluation of solar cooker made from locally sourced materials. [IJMPERD]. *International Journal of Mechanical and Production Engineering Research and Development*, *9*(2), 299–310.

Aemro, Y., Moura, P., & de Almeida, A. (2021). Inefficient cooking systems a challenge for sustainable development: A case of rural areas of Sub-Saharan Africa. *Environment, Development and Sustainability*, *23*, 14697. https://link.springer.com/article/10.1007/s10668-021-01266-7

Ahamed, S., Awalin, L. J., Hasan, S. M. N., Saha, A., Ali, M. S., Syafiq, A., & Wang, L. (2024). Quasi resonant topology based highly efficient solar-powered induction cooker. Bulletin of Electrical Engineering and Informatics, 13(6), 4506-4521.

AKOY, Elamin OM & AHMED, Abdalla IA. (2015). Design, construction and performance evaluation of solar cookers. *Journal of Agricultural Science and Engineering*, *1*(2), 75–82.

Albalak R, Bruce N, McCracken JP, Smith KR, De Gallardo T (2001). Indoor respirable particulate matter concentrations from an open fire, improved cookstove, and LPG/open fire combination in a rural Guatemalan community. Environ Sci Technol. Jul 1;35(13):2650-5. . PMID: 11452588.DOI: 10.1021/es001940m

Algifri, A. H., & Al-Towaie, H. A. (2001). Efficient orientation impacts of box-type solar cooker on the cooker performance. *Solar Energy*, *70*(2), 165–170. DOI: 10.1016/s0038-092x(00)00136-5

Ali, A. O., Elgohr, A. T., El-Mahdy, M. H., Zohir, H. M., Emam, A. Z., Mostafa, M. G., & Elhadidy, M. S. (2025). Advancements in photovoltaic technology: A comprehensive review of recent advances and future prospects. *Energy Conversion and Management*, *X*, 100952.

AL-NEHARI, Hamoud A., MOHAMMED, Mahmoud A., ODHAH, Abdulkarem A., K.A. Al-attab., Bakal K. Mohammed, Abdulwahab M. Al-Habari, Nasr H. Al-Fahd. (2021). Experimental and numerical analysis of tiltable box-type solar cooker with tracking mechanism. *Renewable Energy*, *180*, 954–965.

AMER, Emad H. (2003). Theoretical and experimental assessment of a double exposure solar cooker. *Energy Conversion and Management*, *44*(16), 2651–2663.

American Society of Agricultural and Biological Engineers. (2013). *ASAE Standard S-580.1 Testing and Reporting Solar Cooker Performance.* Standard, ASABE St. Joseph, Michigan, USA.

ANILKUMAR, B. C., MANIYERI, Ranjith, et ANISH, S. (2021). Optimum selection of phase change material for solar box cooker integrated with thermal energy storage unit using multi-criteria decision-making technique. *Journal of Energy Storage*, *40*, 102807.

Ansari, S., Ayob, A., Lipu, M. S. H., Saad, M. H. M., & Hussain, A. (2021). A review of monitoring technologies for solar PV systems using data processing modules and transmission protocols: Progress, challenges and prospects. *Sustainability*, *13*(15), 8120.

Aquilanti, A., Tomassetti, S., Coccia, G., Muccioli, M., & Di Nicola, G. (2023, March). Experimental characterization and performance comparison of four prototypes of panel solar cooker for low to high sun elevations. *Journal of Cleaner Production*, *390*, 136158. DOI: 10.1016/j.jclepro.2023.136158

Aramesh, M., Ghalebani, M., Kasaeian, A., Zamani, H., Lorenzini, G., & Mahian, O.. (2019). A review of recent advances in solar cooking technology. *Renewable Energy*, *140*, 419–435. DOI: 10.1016/j.renene.2019.03.021

Arun Sampaul Thomas, G., Muthukaruppasamy, S., Saravanan, K., & Muleta, N. (2023). Revolutionizing Smart Grids with Big Data Analytics: A Case Study on Integrating Renewable Energy and Predicting Faults. In *Data Analytics for Smart Grids Applications—A Key to Smart City Development* (pp. 179–198). Springer Nature Switzerland.

Arunachala, U., & Kundapur, A. (2020). Cost-effective solar cookers: A global review. *Solar Energy*, *207*, 903–916. DOI: 10.1016/j.solener.2020.07.026

Ashok, K. & Sudhir, C.V. (2009). Proposal for new world standard for testing solar cookers. *Journal of Engineering Science and Technology*. Vol. 4, N°. 3. pp.272 – 281. School of Engineering, Taylor's University College. India.

Atmane, I., Kassmi, V., Deblecker, V., & Bachiri, V. (2021). Realization of autonomous heating plates operating with photovoltaic energy and solar batteries. *Materials Today: Proceedings*, *45*(8), 7408–7414. DOI: 10.1016/j.matpr.2021.01.604

Attou, Y. (2019). Numerical and Thermal Study of a Turbulent Flow in the Annular Space of a Rotor-Stator. Doctoral dissertation, University of Oran 2, Mohamed Ben Ahmed.

Bailis, R., Drigo, R. A. G., & Masera, O. (2015). The carbon footprint of traditional. *Nature Climate Change*, *5*, 266–272.

Balachandran, S., & Swaminathan, J. (2022). Advances in Indoor Cooking Using Solar Energy with Phase Change Material Storage Systems. *Energies*, *15*, 8775. DOI: 10.3390/en15228775

Batchelor, S., Talukder, M. A. R., Uddin, M. R., Mondal, S. K., Islam, S., Redoy, R. K., & Khan, M. R. (2018). Solar e-cooking: A proposition for solar home system integrated clean cooking. *Energies*, *11*(11), 2933.

Bejan, A. (1987). Unification of three different theories concerning the ideal conversion of enclosed radiation. *Journal of Solar Energy Engineering*, *109*(1), 46–51. DOI: 10.1115/1.3268177

Bhave, A. G., & Kale, C. K. (2020). Development of a thermal storage type solar cooker for high temperature cooking using solar salt. *Solar Energy Materials and Solar Cells*, *208*, 110394. DOI: 10.1016/j.solmat.2020.110394

Bhave, A. G., & Thakare, K. A. (2018). Development of a solar thermal storage cum cooking device using salt hydrate. *Solar Energy*, *171*, 784–789. DOI: 10.1016/j.solener.2018.07.018

Biran, A., Abbot, J., & Mace, R. (2004). Families and Firewood: A Comparative Analysis of the Costs and Benefits of Children in Firewood Collection and Use in Two Rural Communities in Sub-Saharan Africa. *Human Ecology*, *32*, 1–25. DOI: 10.1023/B:HUEC.0000015210.89170.4e

BishwashPaneru. (2024). BiplovPaneru, Vikram Alexander, Silvia Nova, Nawraj Bhattarai, RamhariPoudyal, Khem Narayan Poudyal, Mohan B. Dangi, John J. Boland(2024). Solar energy for operating solar cookers as a clean cooking technology in South Asia: A review. *Solar Energy*, *283*, 113004. DOI: 10.1016/j.solener.2024.113004

Bouhamida, M., & Henichi, I. (2020). *Etude et conception d'un cuiseur solaire à réflecteur composé avec tube sous vide*. Université de Ghardaïa.

Bruce, N., Perez-Padilla, R., & Albalak, R. (2000). Indoor air pollution in developing countries:a major environmental and public health challenge. Bulletin of the World Health Organization, *78*(9), 1078. Retrieved 01 08, 2024, from https://pubmed.ncbi.nlm.nih.gov/11019457/

Bureau of Indian Standards (BIS). (1992). *Indian Standard-Solar Cooker, IS 13429.* Manak Bhawan.

Bureau of Indian Standards (BIS). (2000). *Indian standards IS 13429: solar cooker e box type, first revision.* BIS, Manak Bhawan.

Cardoso, J., Couto, A., Costa, P., Rodrigues, C., Facão, J., Loureiro, D., & Simões, T. (2023). Solar Resource and Energy Demand for Autonomous Solar Cooking Photovoltaic Systems in Kenya and Rwanda. *Solar, 3*, 487-503. https://www.sciencedirect.com/science/article/pii/S0038092X21002280

Channiwala, S. A., & Doshi, N. I. (1989). Heat loss coefficients for box-type solar cookers. *Solar Energy*, *42*(6), 495–501.

Coccia, G., Aquilanti, A., Tomassetti, S., Ishibashi, A., & Di Nicola, G. (2021). Design, manufacture and test of a low-cost solar cooker with high-performance light-concentrating lens. *Solar Energy*, *224*, 1028–1039. DOI: 10.1016/j.solener.2021.06.025

Coccia, G., Di Nicola, G., Pierantozzi, M., Tomassetti, S., & Aquilanti, A. (2017). Design, manufacturing, and test of a high concentration ratio solar box cooker with multiple reflectors. *Solar Energy, 155*, 781–792. DOI: 10.1016/j.solener.2017.07.020

Cuce, E., & Cuce, P. M. (2013). A comprehensive review on solar cookers. *Applied Energy, 102*, 1399–1421. DOI: 10.1016/j.apenergy.2012.09.002

Cuce, P. M. (2018). Box type solar cookers with sensible thermal energy storage medium: A comparative experimental investigation and thermodynamic analysis. *Solar Energy, 166*, 432–440. DOI: 10.1016/j.solener.2018.03.077

Damayanti, R. W., Setiadi, H., Laksono, P. W., Rizky, D. L., & Entifar, N. A. (2023). Factors affecting technological readiness and acceptance of induction stoves: A pilot project. *Emerging Science Journal, 7*(6), 1892–1923.

Das, I. a. (2025). The Costs and Benefits of Clean Cooking Policies in Low- and Middle-Income Countries Under Real-World Conditions. *Sustainable Development, n. a.*(n. a.), n. a. DOI: 10.1002/sd.3463

Das, A., & Jain, R. K. (2025). Performance evaluation of mobile robotic system for monitoring and inspection using MQTT protocol and IoT technique. *International Journal of Intelligent Robotics and Applications*, (June). Advance online publication. DOI: 10.1007/s41315-025-00465-z

De, D., Shawhatsu, N., De, N., & Ajaero, M. (2013). Energy-efficient cooking methods. *Energy Efficiency, 6*, 163–175. https://link.springer.com/article/10.1007/s12053-012-9173-7

Delapedra Silva, V., Ferreira, P., Cunha, J., & Kimura, H. (2022). Methods for financial assessment of renewable energy projects: A review. *Processes (Basel, Switzerland), 10*(2), 184.

Dessart, F. J., Barreiro-Hurlé, J., & Van Bavel, R. (2019). Behavioural factors affecting the adoption of sustainable farming practices: A policy-oriented review. *European Review of Agriculture Economics, 46*(3), 417–471.

Djebara A. and Guettaf Z. (2018). Etude comparative entre deux cuiseurs solaires testés sous conditions climatiques de Ghardaïa. 2018-2019.

Djeghab A. and Bougoffa E. (2022). Etude expérimentale d'un cuiseur solaire à faible cout pour les applications domestiques, Master Sciences et Techniques, Spécialité Energie renouvelables en Mécanique, 2022-2023.

Duffie, J. A., & Beckman, W. A. (2013). *Solar Engineering of Thermal Processes* (4th ed.). John Wiley & Sons Ltd.

Duffie, J. A., & Beckman, W. A. (2013). *Solar engineering of thermal processes* (4th ed.). John Wiley and Sons.

Ekechukwu, O. V., & Ugwuoke, N. T. (2003). Design and measured performance of a plane reflector augmented box-type solar-energy cooker. *Renewable Energy*, *28*(12), 1935–1952.

El Moussaoui, N., Talbi, S., Atmane, I., Kassmi, K., Schwarzer, K., & Chayeb, H.. (2020). Feasibility of a new design of a Parabolic Trough Solar Thermal Cooker (PSTC). *Solar Energy*, *201*, 866–871. DOI: 10.1016/j.solener.2020.03.079

El-Sebaii, A. A., & Ibrahim, A. (2005). Experimental testing of a box-type solar cooker using the standard procedure of cooking power. *Renewable Energy*, *30*(12), 1861–1871. DOI: 10.1016/j.renene.2005.01.007

Engoor, G. G. S. S, and Veerappan AR. (2025). Experimental investigation of a box-type solar cooker incorporated with Fresnel lens magnifier. *Energy Sources. Part A, Recovery, Utilization, and Environmental Effects*, *47*, 496–511. DOI: 10.1080/15567036.2020.1826009

Farmer N, Cotter EW (2021). Well-Being and Cooking Behavior: Using the Positive Emotion, Engagement, Relationships, Meaning, and Accomplishment (PERMA) Model as a Theoretical Framework. Front Psychol. Apr 12;12:560578. doi: .DOI: 10.3389/fpsyg.2021.560578

Farooqui, S. Z. (2013). A gravity based tracking system for box type solar cookers. *Solar Energy*, *92*, 62–68. DOI: 10.1016/j.solener.2013.02.024

Felix, P. G., Rajagopal, V., & Kumaresan, K. (2025). Characterisation of erythritol as a potential phase change material. *Australian Journal of Mechanical Engineering*, *23*(2), 205–219.

Funk, P. (2000). Evaluation the international standard procedures for testing solar cookers and reporting performance. *Solar Energy*, *68*(11), 1–7.

Funk, P. A. (2000). Evaluating the international standard procedure for testing solar cookers and reporting performance. *Solar Energy*, *68*(1), 1–7.

Funk, P. A., & Larson, D. L. (1998). Parametric model of solar cooker performance. *Solar Energy*, *62*(1), 63–68. DOI: 10.1016/S0038-092X(97)00074-1

Gama, , A. et al. (2017). Détermination des performances d'un cuiseur solaire boite suivant les procédures de Mullik et Funk. *Rev. Roum. Sci. Techn.* – [Bucarest.]. *Electrotechnique et Energie*, *62*(2), 157–164.

Garg, H. P., Mullick, S. C., & Bhargava, A. K. (1985). Sensible Heat Storage. *Sol Therm Energy Storage*, *7*, 82–153. DOI: 10.1007/978-94-009-5301-7_2

Geddam, S., Dinesh, G. K., & Sivasankar, T. (2015). Determination of thermal performance of a box type solar cooker. *Solar Energy*, *113*, 324–331. DOI: 10.1016/j.solener.2015.01.014

Geng, X., & Bai, L. (2023, October). Characteristics of particulate matter and polycyclic aromatic hydrocarbon pollution generated during kitchen cooking and health risk assessment. *Indoor and Built Environmen, n.a.*(n.a.), n.a. DOI: 10.1177/1420326X231219999

Ghosh, S. S., Biswas, P. K., & Neogi, S. (2017). Thermal performance of solar cooker with special cover glass of low-e antimony doped indium oxide (IAO) coating. *Applied Thermal Engineering*, *113*, 103–111. DOI: 10.1016/j.applthermaleng.2016.10.185

Gupta, P. K., Misal, A., & Agrawal, S. (2021). Development of low cost reflective panel solar cooker. *Materials Today: Proceedings*, *45*, 3010–3013. DOI: 10.1016/j.matpr.2020.12.004

H. Zeghida *et al*, "Enhancing IoT cyber attacks intrusion detection through GAN-based data augmentation and hybrid deep learning models for MQTT network protocol cyber attacks", *Cluster Comput*, vol. 28, n(o) 1, p. 58, Nov. 2024, doi: .DOI: 10.1007/s10586-024-04752-5

H. Zeghida, M. Boulaiche, R. Chikh, A. Patel, A. L. B. Barros, and A. M. Bamhdi, "XMID-MQTT: explaining machine learning-based intrusion detection system for MQTT protocol in IoT environment," *Int. J. Inf. Secur.* vol. 24, n(o) 3, p. 128, May 2025, doi: .DOI: 10.1007/s10207-025-01036-w

Harmim, A., Belhamel, M., Boukar, M., & Amar, M. (2010). Experimental investigation of a box-type solar cooker with a finned absorber plate. *Energy*, *35*(9), 3799–3802. DOI: 10.1016/j.energy.2010.05.032

Harmim, A., Merzouk, M., & Boukar, M. M. (2013). Design and experimental testing of an innovative building-integrated box type solar cooker. *Solar Energy*, *98*(C), 422–433. DOI: 10.1016/j.solener.2013.09.019

Harmim, A., Merzouk, M., Boukar, M., & Amar, M. (2012). Performance study of a box-type solar cooker employing an asymmetric compound parabolic concentrator. *Energy*, *47*(1), 471–480.

Harmin, A., Boukar, M., & Amar, M. (2007). Augmentation du gain énergétique d'un cuiseur solaire à double exposition. *Colloque International sur les Energies Renouvelables CER07, 4-5 Mai 2007*, Oujda. Maroc.

Harmin, A., Boukar, M., & Amar, M. (2007). Etude expérimentale d'un cuiseur solaire de type boite à trois reflecteurs plans et une surface d'ouverture inclinée. *Revue des Énergies Renouvelables*, *10*(11), 31–38.

Hernández-Luna, G., & Huelsz, G. (2008). A solar oven for intertropical zones: Evaluation of the cooking process. *Energy Conversion and Management*, *49*(12), 3622–3626. DOI: 10.1016/j.enconman.2008.07.012

Heywood, H. (1971). Operating experiences with solar water heating. *Journal of the Institute of Heating and Ventilating Engineers*, *39*, 63–69. DOI: 10.4236/oalib.1105445

Holman, J. (2011). *Experimental methods for engineers* (18th ed.). McGraw-Hill Education.

Hosseinzadeh, M., Sadeghirad, R., Zamani, H., Kianifar, A., Mirzababaee, S. M., & Faezian, A. (2021). Experimental study of a nanofluid-based indirect solar cooker: Energy and exergy analyses. *Solar Energy Materials and Solar Cells*, *221*, 110879. DOI: 10.1016/j.solmat.2020.110879

IBRAHIM, Said MA., & EL-REIDY, Medhat K. (2007). The performance of a solar cooker in Egypt. *Renewable Energy*, *6*(8), 1041–1050.

Imani, G., & Moore-Delate, E. (n.d.)."Rapport d'étude de la consommation de bois-énergie et des équipements de cuisson de la ville de Kisangani", [Online]. Available at: https://www.cifor-icraf.org/publications/pdf_files/Reports/Imani-GML-Report.pdf

Incropera, F., DeWitt, D., Bergman, T., & Invine, A. (2007). *Fundamentals of Heat and Mass Transfer* (6th ed.). John Wiley.

ISHAQ, Muhammad., OLUWOLE. Fasiu Ajani., AMULAH. Nuhu Caleb., Bello Abdulsalam. (2023). Performance Evaluation of Box-Type Solar Cookers Using Different Insulation Materials. Nigerian Journal of Science and Engineering Infrastructure vol. 1, no 1.

ISO. (1995). *BIPM, IEC, IFCC, ISO, IUPAC, IUPAP, OIML. Guide to the Expression of Uncertainty in Measurement*. International Standards Office.

Jebaraj, S., & Srinivasa, P. R. (2015). High-efficiency solar oven for tropical countries. *Journal of Engineering and Applied Sciences (Asian Research Publishing Network)*, *10*, 10213–10217.

Jones, W. Peter, LAUNDER, Brian Edward. (1972). The prediction of laminarization with a two-equation model of turbulence. *International Journal of Heat and Mass Transfer*, *15*(2), 301–314.

Joshi, S. B., & Jani, A. R. (2015). Design, development and testing of a small scale hybrid solar cooker. *Solar Energy*, *122*, 148–155.

Jubran, B. A., & Alsaad, M. A. (1991). Parametric study of a box-type solar cooker. *Energy Conversion and Management*, *32*(3), 223–234.

Kassmi, K. (2021). Autonomous solar cooker with photovoltaic energy. *New Journal of Science Vulgarization.*, *1*(2), 54–62.

Kaushik, S., & Gupta, M. (2008). Energy and exergy efficiency comparison of community-size and domestic-size paraboloidal solar cooker performance. *Energy for Sustainable Development*, *12*(3), 60–64. DOI: 10.1016/S0973-0826(08)60440-8

KHALDI. Souheyla Yahiaoui. (2018). Numerical Study of the Thermal Behavior of a Solar Dryer Using a Thermal Bed for Energy Storage (2018). Doctoral dissertation, University of Bourgogne Franche-Comté; Abou Bekr Belkaid University (Tlemcen, Algeria).

Khalifa, Y., & Al, E. (2025). The role of electric cooking in providing sustainable school meals in low-income and lower-middle-income countries. *The Lancet. Planetary Health*. Advance online publication. DOI: 10.1016/S2542-5196(25)00004-X PMID: 40054485

Khennoufa A., Lachibi O. (2017). *Etude d'un cuiseur solaire*. 2017-2018.

Kimambo, C. (2007). Development and performance testing of solar cookers. *Journal of Energy in Southern Africa*, *18*, 41–51. DOI: 10.17159/2413-3051/2007/v18i3a3384

Kinyua, V., Wambugu, A., & da Silva, I. (2025). *D10.8 Report on field testing of standalone solar cooking appliance. European Union's Horizon 2020 Research and Innovation Program under Grant Agreement 963530., Long-Term Joint EU-AU Research and Innovation Partnership on Renewable Energy LEAP-RE. WP10 Task 10.2*. Strathmore University., Retrieved from https://www.leap-re.eu/wp-content/uploads/2025/02/LEAP-RE-D10.8-Report-on-field-testing-of-standalone-solar-cooking-appliance-V2.pdf

Koh, K., Kim, Y.-J., & Kim, H.-S. (2016). Nutrition Analyses of Nutrition Bar Cooked with a Solar Box-cooker. *Korean Journal of Human Ecology.*, *25*, 487. DOI: 10.5934/kjhe.2016.25.4.487

Kotas, T. J. (1985). Basic exergy concepts. In *The Exergy Method of Thermal Plant Analysis* (pp. 29–56). https://doi.org/DOI: 10.1016/b978-0-408-01350-5.50009-x

Krane, R. J. (1987). A second law analysis of the optimum design and operation of thermal energy storage systems. *International Journal of Heat and Mass Transfer*, *30*(1), 43–57. DOI: 10.1016/0017-9310(87)90059-7

KREITH. F., KREIDER, J. F., & GOSWAMI, D. Y. (2000). *Principles of solar engineering*. Taylor & Francis.

KULDHAR. Santosh & SHRIVASTAVA, Ritu. (2023). Development and Fabrication of Box Type Solar cooker with PCM to Enhance Its Efficiency, vol 10.

Kumar Goyal, R., & Eswaramoorthy, M. (2024). Theoretical and experimental analysis of box-type solar cooker with sensible heat storage. *Solar Energy*, *268*, 112273. DOI: 10.1016/j.solener.2023.112273

KUMAR, Naveen, AGRAVAT, Sagar, CHAVDA, H.N. Mistry. (2008). Design and development of efficient multipurpose domestic solar cookers/dryers. *Renewable Energy*, *33*(10), 2207–2211.

Kumar, A., Saxena, A., Pandey, S. D., & Gupta, A. (2023). Cooking performance assessment of a phase change material integrated hot box cooker. *Environmental Science and Pollution Research International*, *31*, 62392–62407. DOI: 10.1007/s11356-023-25340-x PMID: 36689111

Kumar, N., Chavda, T., & Mistry, H. N. (2010). A truncated pyramid non-tracking type multipurpose domestic solar cooker/hot water system. *Applied Energy*, *87*(2), 471–477. DOI: 10.1016/j.apenergy.2009.06.031

Kumar, S. (2004). Thermal performance study of box type solar cooker from heating characteristic curves. *Energy Conversion and Management*, *45*(1), 127–139. DOI: 10.1016/s0196-8904(03)00103-1

Kundapur, A. (2018). A treatise on Solar Cookers. [Kalashree, Hayagreva Nagara: International Alternate energy Trust.]. *Udupi*, *576*, 102.

Kundapur, A., & Sudhir, C. V. (2009). Proposal for a new world standard for testing solar cookers. *Journal of Engineering Science and Technology*, *4*(3), 272–281.

Lahkar, P., & Samdarshi, S. (2010). A review of the thermal performance parameters of box type solar cookers and identification of their correlations. *Renewable & Sustainable Energy Reviews*, *14*(6), 1615–1621. DOI: 10.1016/j.rser.2010.02.009

Lakshminarayana, S., & Santhi Thilagam, P. (2025, June). Next-Generation DDoS Attacks on IoT Deployments: Targeting the Advanced Features of MQTT v5.0 Protocol. *IEEE Internet of Things Journal*, *12*(12), 22090–22109. DOI: 10.1109/JIOT.2025.3549784

Leach, M., Mullen, C., Lee, J., Soltowski, B., Wade, N., Galloway, S., & Batchelor, S. (2021). Modelling the costs and benefits of modern energy cooking services—Methods and case studies. *Energies, 14*(12), 3371.

Leary, J., Leach, M., Batchelor, S., Scott, N., & Brown, E. (2021). Battery-supported eCooking: A transformative opportunity for 2.6 billion people who still cook with biomass. *Energy Policy, 159*, 112619.

Lecuona, A., Nogueira, J. I., Ventas, R., Rodríguez-Hidalgo, M. C., & Legrand, M. (2003). Solar cooker of the portable parabolic type incorporating heat storage based on PCM. *Applied Energy, 111*, 1136–1146. DOI: 10.1016/j.apenergy.2013.01.083

Lecuona-Neumann, A., Nogueira, J., & Legrand, M. (2018). Photovoltaic Cooking, Chapter 13. In I. Yahyaoui, *Advances in Renewable Energies and Power Technologies. Volume 1: Solar and Wind Energies*. Elsevier. DOI: 10.1016/C2016-0-04518-7

Lecuona-Neumann, A. (2017). *Cocinas solares. Fundamentos y aplicaciones. Herramientas de lucha contra la pobreza energética*. Editorial Marcombo.

Lecuona-Neumann, A., Nogueira-Goriba, J., Famiglietti, A., & Rodríguez-Hidalgo, M., & Boubour. (2024). Photovoltaic Cooker with No Electronics or Battery. *Energies, 17*, 1192. DOI: 10.3390/en17051192

Liu, W., Luo, X., Huang, Y., Zhao, M., Liu, T., Wang, J., & Feng, F. (2023). Influence of cooking techniques on food quality, digestibility, and health risks regarding lipid oxidation. *Food Research International, 167*, 112685. DOI: 10.1016/j.foodres.2023.112685 PMID: 37087258

Łosiewicz, B. (2024). Recent Advances in Technologies of Green Hydrogen Production. *Energies, 17*, 4514.

Lugolole, R., Mawire, A., Lentswe, K., Okello, D., & Nyeinga, K. (2018). Thermal performance comparison of three sensible heat thermal energy storage systems during charging cycles. *Sustainable Energy Technologies and Assessments*, 37–51. DOI: 10.1016/j.seta.2018.09.002

Mahavar, S., Sengar, N., Rajawat, P., Verma, M., & Dashora, P. (2012). Design development and performance studies of a novel single family solar cooker. *Renewable Energy, 47*, 67–76. DOI: 10.1016/j.renene.2012.04.013

Marín, A. I., Abdul Malak, D., Bastrup-Birk, A., Chirici, G., Barbati, A., & Kleeschulte, S. (2021, September). Mapping forest condition in Europe: Methodological developments in support to forest biodiversity assessments. *Ecological Indicators, 128*, 107839. DOI: 10.1016/j.ecolind.2021.107839

Mawire, A., & Abedigamba, O. (2025). Experimental comparison of erythritol and erythritol-granite pebble mixtures as heat storage materials for solar cooking. Green Technology & Innovation. *Green Technology & Innovation, 1*(1), 6191. DOI: 10.36922/gti.6191

Mawire, A., Abedigamba, O. P., & Worall, M. (2024, February). Experimental comparison of a DC PV cooker and a parabolic dish solar cooker under variable solar radiation conditions. *Case Studies in Thermal Engineering, 54*, 103976. DOI: 10.1016/j.csite.2024.103976

Mawire, A., Lentswe, K., & Owusu, P. (2022). Performance of two solar cooking storage pots using parabolic dish solar concentrators during solar and storage cooking periods with different heating loads. *Results in Engineering, 13*, 100336. DOI: 10.1016/j.rineng.2022.100336

Mbodji, N., & Hajji, A. (2017). Analyse technico-économique et évaluation de l'impact environnemental de la cuisson solaire directe au Maroc. Rev. Mar. sci. Agron. *Vét.*, *5*(3), 325–332.

Mehta K, Thrikutam N, Hoyte-Williams PE, Falk H, Nakarmi K, Stewart B. (2023) Epidemiology and Outcomes of Cooking- and Cookstove-Related Burn Injuries: A World Health Organization Global Burn Registry Report. J Burn Care Res. May 2;44(3):508-516. doi: .DOI: 10.1093/jbcr/irab166

Mirdha, U. S., & Dhariwal, S. R. (2008). Design optimization of solar cooker. *Renewable Energy, 33*(3), 530–544. DOI: 10.1016/j.renene.2007.04.009

Mishra, R. S., & Sabberwal Prakash, S. P. (1984). Evaluation of solar cooker thermal performance using different insulating materials. *International Journal of Energy Research, 8*(4), 393–396. DOI: 10.1002/er.4440080410

Misra, N., Anand, A., Pandey, S., Kant, K., Shukla, A., & Sharma, A. (2023). Box-type solar cookers: An overview of technological advancement, energy, environmental, and economic benefits. *Energies, 16*(4), 1697.

Misra, R., & Aseri, T. K. (2012). Thermal performance enhancement of box-type solar cooker: A new approach. *International Journal of Sustainable Energy, 31*(2), 107–118. DOI: 10.1080/1478646x.2011.552978

Mitali, J., Dhinakaran, S., & Mohamad, A. A. (2021). Energy storage systems: A review. *Energy Storage and Saving, 1*(3), 166–216.

Mosbah, A. & Achour, A. (2020). Contribution à la détermination expérimentale des performances d'un cuiseur solaire testé sous climat aride du Sude Algérien, master Energies Renouvelables en Electrotechnique, Université de Ghardaia, Faculté des Sciences et Technologies Département d'Automatique et Electromécaniques, 2020-2021.

MUJAHID, A. M. (1994). Optimum tilt angle for solar collection systems. *International Journal of Solar Energy*, *14*(4), 191–202. DOI: 10.1080/01425919408909810

Mukelabai, M. D., Wijayantha, K. G. U., & Blanchard, R. E. (2022). Hydrogen for cooking: A review of cooking technologies, renewable hydrogen systems and techno-economics. *Sustainability*, *14*(24), 16964.

Mullick, S. C., Kandpal, T. C., & Saxena, A. K. (1987). Thermal test procedure for box-type solar cookers. *Solar Energy*, *39*(4), 353–360. DOI: 10.1016/s0038-092x(87)80021-x

Mullick, S. C., Kandpal, T. C., & Saxsena, A. K. (1987). Thermal test procedure for box type solar cookers. *Solar Energy*, *39*, 353–360.

Mullick, S., Kumar, T., & Kumar, S. (1996). Testing of box type solar cooker: Second figure of merit F2 and its variation with load and number of pots. *Solar Energy*, *57*(15), 409–413.

Nahar, N. (2003). Performance and testing of a hot box storage solar cooker. *Energy Conversion and Management*, *44*, 1323–1331. DOI: 10.1016/S0196-8904(02)00113-9

Nahar, N. (2003). Performances and testing of a hot box storage solar cooker. *Energy Conversion and Management*, *44*, 1323–1331.

Nahar, N. M. (1990). Performance and testing of an improved hot box solar cooker. *Energy Conversion and Management*, *30*(1), 9–16. DOI: 10.1016/0196-8904(90)90051-y

Nahar, N. M. (1998). Design, development and testing of a novel non-tracking solar cooker. *International Journal of Energy Research*, *22*, 1191–1198.

Nahar, N. M., Marshall, R. H., & Brinkworth, B. J. (1994). Studies on a hot box solar cooker with transparent insulation materials. *Energy Conversion and Management*, *35*(9), 787–791. DOI: 10.1016/0196-8904(94)90107-4

Nandha Gopal, J., Muthukaruppasamy, S., Rajagopal, V., Thomas, G. A. S., & Parimalarsundar, E. (2024). Artificial Intelligence Governance and Comprehensibility in Renewable Energy Systems. In Explainable AI (XAI) for Sustainable Development (pp. 73-88). Chapman and Hall/CRC.

Nandwani, S. S. (1996). Solar cookers – cheap technology with high ecological benefits. *Ecological Economics*, *17*, 73–81.

Ning, Ke; Zhou, Yan-Yan; Zhang, Ning; Sun, Xue-Jun; Liu, Wen-Wu; Han, Cui-Hong MD (2020). Neurocognitive sequelae after carbon monoxide poisoning and hyperbaric oxygen therapy. Medical Gas Research 10(1):p 30-36, Jan–Mar. | DOI: DOI: 10.4103/2045-9912.279981

Nkhonjera, L., Bello-Ochende, T., John, G., & King'ondu, C. K. (2017). A review of thermal energy storage designs, heat storage materials and cooking performance of solar cookers with heat storage. *Renewable & Sustainable Energy Reviews*, *75*, 157–167. DOI: 10.1016/j.rser.2016.10.059

Obayedullah, M. (2025). Review Article Recent Advances in Phase Change Energy Storage Materials. *Developments and Applications*, *2025*. Advance online publication. DOI: 10.1155/er/6668430

Odoi-Yorke, F., Baah, B., & Opoku, R. (2025). Scientific mapping of experimental research on solar cookers: Global trends, evolution, and future directions. *Solar Energy Advances*, *5*, 100093. DOI: 10.1016/j.seja.2025.100093

Ossei-Bremanga, R. N., & Al., E. (2024). Choice modelling of eCooking adoption by households in Ghana. *International Journal of Sustainable Engineering*, *17*(1), 1–13. DOI: 10.1080/19397038.2024.2327832

Ozturk, H. H. (2004). Energy and exergy efficiencies of a solar box-cooker. *International Journal of Exergy*, *1*(2), 202. DOI: 10.1504/ijex.2004.005091

Padonou, E. A., Akabassi, G. C., Akakpo, B. A., & Sinsin, B. (2022). Importance of solar cookers in women's daily lives: A review. *Energy for Sustainable Development*, *70*, 466–474. DOI: 10.1016/j.esd.2022.08.015

Panagoda, L. P. S. S., Sandeepa, R. A. H. T., Perera, W. A. V. T., Sandunika, D. M. I., Siriwardhana, S. M. G. T., Alwis, M. K. S. D., & Dilka, S. H. S. (2023). Advancements in photovoltaic (Pv) technology for solar energy generation. Journal of research technology & engineering, 4(30), 30-72.

Pandey, A. K., Tyagi, V. V., Park, S. R., & Tyagi, S. K. (2012). Comparative experimental study of solar cookers using exergy analysis. *Journal of Thermal Analysis and Calorimetry*, *109*, 425–431.

Paneru, B., Paneru, B., Alexander, V., Nova, S., Bhattarai, N., Poudyal, R., . . . Boland, J. J. (2024). Solar energy for operating solar cookers as a clean cooking technology in South Asia: A review. Solar Energy, 283, 113004. Chiloane, L. L., Aswat, M., Yen, Y. C. J., & Cronje, W. A. (2023). A PV-supplied cooking solution using a hybrid electric-thermal energy storage system. In 2023 31st Southern African Universities Power Engineering Conference (SAUPEC) (pp. 1-5). IEEE.

Panwar, N. L., Kaushik, S. C., & Kothari, S. (2011). Role of renewable energy sources in environmental protection: A review. *Renewable & Sustainable Energy Reviews*, *15*, 1513–1524.

Panwar, N. L., Kothari, S., & Kaushik, S. C. (2010). Experimental investigation of energy and exergy efficiency of masonry-type solar cooker for animal feed. *International Journal of Sustainable Energy*, *29*(3), 178–184. DOI: 10.1080/14786461003628679

Panwar, N., Kaushik, S., & Kothari, S. (2012). State of the art of solar cooking: An overview. *Renewable & Sustainable Energy Reviews*, *16*(6), 3776–3785.

Patel, N., & Philip, S. (2000). Performance evaluation of three solar concentrating cookers. *Renewable Energy*, •••, 347–355.

Petela, R. (2003). Exergy of undiluted thermal radiation. *Solar Energy*, *74*(6), 469–488. DOI: 10.1016/s0038-092x(03)00226-3

Prasanna, U. R., & Umanand, L. (2011). Optimization and design of energy transport system for solar cooking application. *Applied Energy*, *88*(1), 242–251.

Rahul Kashyap, S., Pramanik, S., & Ravikrishna, R. (2023). A review of solar, electric and hybrid cookstoves. *Renewable & Sustainable Energy Reviews*, *188*, 113787. DOI: 10.1016/j.rser.2023.113787

Rashid, F. L., Al-Obaidi, M. A., Mahdi, A. J., & Ameen, A. (2024). Advancements in Fresnel lens technology across diverse solar energy applications: A comprehensive review. *Energies*, *17*(3), 569.

Redekar, A., & Deb, D. (2024). Design of an advanced solar cooker with integrated self-powered active sun tracking and dual heating facilitated for rural communities. [IJIDeM]. *International Journal on Interactive Design and Manufacturing*, 1–25.

REH S. (2004). Renewable energy: A response to climate change. *Solar Energy*, *76*, 9–17.

Rehfuess, E. (2006). *Fuel for life: Household energy and health*. World Health Organization.

Rodríguez-Ayala M, Sandoval-Insausti H, Bayán-Bravo A, Banegas JR, Donat-Vargas C, Ortolá R, Rodríguez-Artalejo F, Guallar-Castillón P.(2022) Cooking Methods and Their Relationship with Anthropometrics and Cardiovascular Risk Factors among Older Spanish Adults. Nutrients. Aug 20;14(16):3426. doi: .DOI: 10.3390/nu14163426

Rodríguez-Muñoz, J. D., Tlelo-Cuautle, E., & de la Fraga, L. G. (2025, May). Chaos-based authentication of encrypted images under MQTT for IoT protocol. *Integration (Amsterdam), 102*, 102378. DOI: 10.1016/j.vlsi.2025.102378

Rohsenow, W., Harnett, J., & Cho, Y. (1998). *Handbook of Heat Transfer 3 rd.* McGraw-Hill.

Ronge, H., Niture, V., & Ghodake, M. D. (2016). A review paper on utilization of solar energy for cooking. *Imperial Int J Eco-friendly Technolo, 1*, 121–124.

Rosa, G., Majorin, F., Boisson, S., Barstow, C., Johnson, M., Kirby, M., & Clasen, T. (2014). Assessing the impact of water filters and improved cook stoves on drinking water quality and household air pollution: A randomized controlled trial in Rwanda. *PLoS One, 9*, e91011. Advance online publication. DOI: 10.1371/journal.pone.009101 PMID: 24614750

Sadeq, A. M., Homod, R. Z., Hussein, A. K., Togun, H., Mahmoodi, A., Isleem, H. F., & Moghaddam, A. H. (2024). Hydrogen energy systems: Technologies, trends, and future prospects. *The Science of the Total Environment*, 173622. PMID: 38821273

Saini, D. K., Chandrashekara, M., & Yadav, A. (2025). Thermal analysis of a rectangular latent heat storage unit with stearic acid/paraffin wax composite PCM for solar thermal energy storage systems. *Energy Sources. Part A, Recovery, Utilization, and Environmental Effects, 47*, 4279–4296. DOI: 10.1080/15567036.2025.2460644

Saini, Prashant, Pandey, Sushant, Goswami, Shruti, & Dhar, Atul, M.E. Mohamed, Satvasheel Powar. (2023). Experimental and numerical investigation of a hybrid solar thermal-electric powered cooking oven. *Energy, 280*, 128188. DOI: 10.1016/j.energy.2023.128188

Sarangi, A., Sarangi, A., Sahoo, S., Nayak, J., & Mallik, R. (2024). Advancements and global perspectives in solar cooking technology: A comprehensive review. *Energy Nexus, 13*, 100266. DOI: 10.1016/j.nexus.2023.100266

Sathish. (2025). MWCNT SiO2 Hybrid Nano-PCM for Ultrafast Solar Cookers An Experimental Study (1).pdf n.d.

SAXENA, Abhishek, CUCE, Erdem, TIWARI, G. N., Avnish Kumar. (2020). Design and thermal performance investigation of a box cooker with flexible solar collector tubes: Experimental research. Energy, vol. 206, p. 118144.

Saxena, A., Varun, Pandey, S. P., & Srivastav, G. (2011). A thermodynamic review on solar box type cookers. *Renewable & Sustainable Energy Reviews, 15*(6), 3301–3318. DOI: 10.1016/j.rser.2011.04.017

Saxena, A., Norton, B., Goel, V., & Singh, D. B. (2022). Solar cooking innovations, their appropriateness, and viability. *Environmental Science and Pollution Research International, 29*(39), 58537–58560. PMID: 35778663

Scott, N., Leach, M., & Clements, W. (2024). Energy-Efficient Electric Cooking and Sustainable Energy Transitions. *Energies, 17*(13), 3318.

Selvakumar, P., Muthusamy, S., Satishkumar, D., Vigneshkumar, P., Selvamurugan, C., & Kumar, P. S. (2024). *AI-Powered Tools, IGI Global, Mar*. IGI Global., DOI: 10.4018/979-8-3693-2615-2.ch002

Selvamurugan, C., Delna Renish, M., Ishwarya, F.J., & Hisha Sherin K.(2025). Social Dining Platform:Dining Platforms and their Impact on Food Culture and Community Building" IGI Global, Feb, DOI: .DOI: 10.4018/979-8-3693-9025-2.ch008

Sethi, V. P., Pal, D. S., & Sumathy, K. (2014). Performance evaluation and solar radiation capture of optimally inclined box type solar cooker with parallelepiped cooking vessel design. *Energy Conversion and Management, 81*, 231–241. DOI: 10.1016/j.enconman.2014.02.041

Sharma A, Tyagi V V, Chen CR, Buddhi D. Review on thermal energy storage with phase change materials and applications 2009;13:318–45. https://doi.org/.DOI: 10.1016/j.rser.2007.10.005

Sharma, S. D., Buddhi, D., Sawhney, R. L., & Sharma, A. (2000). Design, development and performance evaluation of a latent heat storage unit for evening cooking in a solar cooker. *Energy Conversion and Management, 41*, 1497–1508. DOI: 10.1016/S0196-8904(99)00193-4

Shukla, S. K. (2009). Comparison of energy and exergy efficiency of community and domestic type parabolic solar cookers. *International Journal of Green Energy, 6*, 437–449.

Sillé, R. Z., & van Duijsen, P. (2023). Low-cost Converter Design to Interface Solar to Inductive Cooking in Rural Areas. *International Conference on Electrical, Computer and Energy Technologies (ICECET)*. Cape Town-South Africa: IEEE. DOI: 10.1109/ICECET58911.2023.10389500

Singh, Sudha, Gupta, Gyan Prakash, & Kumar, Bablu, U.C. Kulshrestha. (2014). Comparative study of indoor air pollution using traditional and improved cooking stoves in rural households of Northern India. *Energy for Sustainable Development, 19*, 1–6. DOI: 10.1016/j.esd.2014.01.007

Soria-Verdugo, A. (2015). Experimental analysis and simulation of the performance of a box-type solar cooker. *Energy for Sustainable Development*, *29*, 65–71. DOI: 10.1016/j.esd.2015.09.006

Swinbank, W. C. (1963). Long-Wave Radiation from Clear Skies. *Quarterly Journal of the Royal Meteorological Society*, *89*, 339–348.

Taşkesen, E., Dirik, M., Tekir, M., & Pazarlioğlu, H. K. (2023). Predicting heat transfer performance of Fe3O4-Cu/water hybrid nanofluid under constant magnetic field using ANN. *J Therm Eng*, *9*, 811–822. DOI: 10.18186/thermal.000000

Terres, H., Lizardi, A., Chávez, S., López, R., & Vaca, M. (2017). Exergy applied to the heat conduction analysis in glass covers of a solar cooker box-type with internal and external reflectors. *Journal of Physics: Conference Series*, *792*, 012012. DOI: 10.1088/1742-6596/792/1/012012

Terres, H., Lizardi, A., López, R., Vaca, M., & Chávez, S. (2014). Mathematical model to study solar cookers box-type with internal reflectors. *Energy Procedia*, *57*, 1583–1592. DOI: 10.1016/j.egypro.2014.10.150

Teske, S., Niklas, S., & Miyake, S. (2023). Technical Scenario for 100% Renewable Energy in Nepal by 2050: Possible Transition Pathways for NDC & LTS Implementation; prepared for World Future Council, WWF and Brot für die Welt by the University of Technology Sydney, Institute for Sustainable Futures; February 2023.

Thirugnanam, C., Karthikeyan, S., & Kalaimurugan, K. (2020). Study of phase change materials and its application in solar cooker. *Materials Today: Proceedings*, *33*, 2890–2896. DOI: 10.1016/j.matpr.2020.02.780

Todd, J. J., & Miller, S. (1999). *Performance testing of cardboard, solar box-cookers.* Graham Thorpe.

Toonen, H. M. (2009). Adapting to an innovation: Solar cooking in the urban households of Ouagadougou (Burkina Faso). *Physics and Chemistry of the Earth Parts A/B/C*, *34*(1–2), 65–71. DOI: 10.1016/j.pce.2008.03.006

TRNSYS. (1979). *A transient simulation program.* Solar Energy Laboratory, University of Wisconsin.

Uhuegbu, C. C. (2011). Design and Construction of a Wooden Solar Box Cooker with performance and efficiency test. *Journal of Basic and Applied Scientific Research, 1*(7), 533–538.

Vaishya, J. S., Tripathi, T. C., Singh, D., Bhawalkar, R. H., & Hegde, M. S. (1985). A hot box solar cooker: Performance analysis and testing. *Energy Conversion and Management, 25*(3), 373–379. DOI: 10.1016/0196-8904(85)90057-3

Venkateshwaran, B. G., Kumaresan, G., Santosh, R., & Velraj, R. (2025). Performance enhancement of flat plate cooking unit using novel hook turbulators for indirect solar cooking applications: Numerical and experimental assessment with enviroeconomic analyses. *Solar Energy, 287*, 113247. DOI: 10.1016/j.solener.2025.113247

Vipin, Singh, N., & Yadav, A. (2025). Fabrication and experimental investigation of a novel sensible heat storage system for indoor solar cooking. *Journal of Energy Storage, 117*, 116201. DOI: 10.1016/j.est.2025.116201

WASSIE. Hailemariam M., GETIE, Muluken Z., ALEM, Mulat S., Kotu Teshome B., (2022). Experimental investigation of the effect of reflectors on thermal performance of box type solar cooker. Heliyon vol. 8, no 12

WASSIE. Hailemariam M., GETIE, Muluken Z., ALEM, Mulat S., *Teshome B., Zelalem M., 2022).* Experimental investigation of the effect of reflectors on thermal performance of box type solar cooker. *Heliyon* vol. 8, no 12.

Wassie, H. M., Getie, M. Z., Alem, M. S., Kotu, T. B., & Salehdress, Z. M. (2022). Experimental investigation of the effect of reflectors on thermal performance of box type solar cooker. *Heliyon, 8*, e12324. DOI: 10.1016/j.heliyon.2022.e12324 PMID: 36582681

Weldu, Amanuel, ZHAO, Li, DENG, Shuai, Mulugeta. Nigussie, Zhang. Ying, Nie. Xianhue, Xu. Weicong. (2019). Performance evaluation on solar box cooker with reflector tracking at optimal angle under Bahir Dar climate. *Solar Energy, 180*, 664–677.

Worls Health Organization (WHO). (2022, January 20). *WHO publishes new global data on the use of clean and polluting fuels for cooking by fuel type*. Retrieved March 3, 2025, from WHO: https://www.who.int/news/item/20-01-2022-who-publishes -new-global-data-on-the-use-of-clean-and-polluting-fuels-for-cooking-by-fuel-type

Yakhlef Y. and Mehdaoui A. (2018). Contribution à la détermination expérimentale des performances thermiques d'un cuiseur solaire parabolique. 2018-2019.

Yayhyaoui, E. (2018). *Poor People's Energy Outlook 2018: Achieving Inclusive Energy Access at Scale*. Practical Action Publishing., DOI: 10.3362/9781780447544

Yettou F. et al. (2012). Etude comparative entre deux types de cuiseurs solaires boites en milieu aride. *Revue des Energies Renouvelables*, Vol. Numéro spécial Sienr'12, pp. 233-245.

Yettou, F. (2015). Conception et réalisation d'un système de cuisson solaire destiné au site saharien (Ghardaïa, Algérie). Thèse de Doctorat en Sciences. Université Hadji Lakhdar Batna Faculté de Technologie Département d'Electrotechnique. Soutenue le 10 Mars 2015.

Yettou, F., Azoui, B., Ali, M., Panwar, N. L., Gama, A., & Toufik, A. (2014). Performance parameters determination by experimental testing of box solar cooker with south tilted collecting surface.

Yettou, F., Azoui, B., Malek, A., Panwar, N. L., & Gama, A. (2015). Energetic and exergetic evaluation of solar box cooker in Algerian climatic conditions. *International Journal of Exergy*, *16*(3), 337. DOI: 10.1504/ijex.2015.068230

Yettou, F., Azoui, B., Malek, A., Panwar, N. L., Gama, A., Arrif, T., & Merarda, H. (2023). Comparative assessment of two different designs of box solar cookers under Algerian Sahara conditions. *Journal of Renewable Energies*, *18*(2). Advance online publication. DOI: 10.54966/jreen.v18i2.500

Yolcan, O.O. (2023). "World energy outlook and state of renewable energy: 10-Year evaluation", *Innovation and Green Development*, vol. 2, n(o) 4, p. 100070, Dec. 2023, doi: .DOI: 10.1016/j.igd.2023.100070

Yu, G., Zhang, G., Poslad, S., Fan, Y., & Xu, X. (2024). A study of quantifying the influence of kitchen human activity on indoor air quality dynamics. *Environmental Pollution*, *362*, 124900. DOI: 10.1016/j.envpol.2024.124900 PMID: 39260554

Zhou, C., Wang, Y., Li, J., Ma, X., Li, Q., & Yang, M.. (2023). Simulation and economic analysis of an innovative indoor solar cooking system with energy storage. *Solar Energy*, *263*, 111816. DOI: 10.1016/j.solener.2023.111816

About the Contributors

Khalil Kassmi, born in Casablanca, Morocco, in 1963, is a distinguished academic and researcher specializing in electronics and renewable energy. He completed his higher education at Paul Sabatier University (PSU) in Toulouse, France, from 1983 to 1991. In 1991, he earned his PhD in Electronics from PSU, conducting research at the Laboratory of Automatic and Analysis of Systems (LAAS/CNRST) within the MOS Technology team. In 1992, he joined SGS Thomson-Casablanca as an engineer, where he was responsible for the assembly line of power electronic components, including TO220 and thyristors. A year later, in 1993, he transitioned to academia, joining Mohammed 1st University in Oujda, Morocco, as a professor-researcher. He obtained his second PhD (Doctorat d'État) in Electronics in 1996. Since 2000, he has been a full professor at the same university. Currently, Khalil Kassmi leads the research team 'Materials, Electronics, and Renewable Energies (MERE),' focusing on materials and electronic components, photovoltaic cells and systems, and solar energy applications. Under his supervision, numerous PhD theses have been successfully completed. His academic contributions include co-authoring over 100 publications and international conference presentations in the field of photovoltaic renewable energy, covering materials, cells, and systems. Professor Kassmi has been coordinating national and international projects related to solar and photovoltaic renewable energy, both in education and research, since 2002 (projects: Actions intégrées, PROTARS, PMARS, LEAP-RE, WBI, ARICA, etc.). He holds five patents in photovoltaic control and regulation, particularly in photovoltaic-powered heating systems. As an expert evaluator for Morocco's National Center for Scientific and Technical Research (CNRST) in Rabat, he has significantly contributed to the scientific community since October 2019. Additionally, he serves as a reviewer for several indexed journals. He is the editor of the book chapter collection Solar Water Heating: Fundamentals and Applications (Nova Science Publishers, 2021) and co-editor of another book chapter collection (Taylor & Francis Group, 2021). His expertise has earned him recognition as a thought leader in renewable energy technologies, and his work continues to inspire advancements in the field

Olivier Deblecker (1971) received the degree in civil electrical engineering and the Ph.D. degree in electrical engineering in 1995 and 2001, respectively. He is currently a Professor and Head of the Electrical Power Engineering Unit at the Faculty of Engineering, University of Mons (Belgium). Since 2018, he is also the academic coordinator of the Joint Certificate Programme in Energy Electronics gathering partners from 4 Belgian universities and two of the major companies in Belgium (Alstom, Thalès Alenia Space) involved in that field. His research activities include the following main topics: power electronic systems and, more particularly, DC-DC power converters for space and auxiliary railway power supply applications (optimal design and control), modelling of electromagnetic systems operating at intermediate frequencies, wireless (contactless) power transfer applications, micro-grids (mainly power quality aspects). He is also a member of EPE - European Power Electronics Association, since 2003, and IEEE - Institute of Electrical and Electronics Engineers, since 2009.

***

Abdennabi Brahmi is Professor in Electrical Engineering Department of Faculty of Sciences and Techniques at Moulay Ismail University, Meknes, Morocco. He is a permanent member in Interdisciplinary Modelling, Simulation and Scientific Innovation Laboratory (LIMSIS) in Moulay Ismail University, Meknes, Morocco. He published more than twenty technical papers in reviews and international conferences. He gives courses for masters students in Electrical Engineering field. And he supervises Master's and Doctoral students in Electrical Engineering and Renewable Energy. He got a Ph.D. diploma in Electrotechnics and Power Electronics from Sultan Moulay Slimane University, Beni-Mellal, Morocco in 2013. Also, he got an Advanced Higher Studies Diploma in Electrotechnics Power Electronics and Industrial Control from Cadi_Ayyad University, Marakech, Morocco in 2006. His research specialization are in Electrotechnics, Power Electronics and Intelligent Control in Renewable Energy and hybrid Systems: Design, Co-Simulation and Realization.

Parimalasundar E obtained Ph.D. degree in Power electronics from Anna University Chennai. He is working at Mohan Babu University, Tirupati, India. His research interests include: Power converters, universal converter and controllers. In the current study he shown proposed methodology and implemented it in the experimental setup.

Ahmed Ihlal was born and brought up in Morocco. He studied Physics and Chemistry and holds, in 1984, his BSc degree (Licence es-Sciences Physique) in Solid State Physics from the University Mohamed V, Rabat - Morocco. He then joined Paris VII University – France, where he got, in 1985, a MSc. degree (DEA: Diplôme des Etudes Approfondies) in Solar Energy. He pursued his research and earned, in 1988, a PhD degree from the University of Caen Basse-Normandie - France. Dr. A. Ihlal started his teaching career on I988 as an Assistant Professor in the faculty of Science at University Ibn Zohr. He holds a "Doctorat d'Etat" thesis in 1995. He is currently Full Professor in Faculty of Sciences, University Ibn Zohr, Agadir - Morocco. He is head of the group working on developing cost effective processes for the fabrication of CIGS and CZTS absorber layers, buffer layers and TCOs. He is working on PV and CSP systems as well. He is developing materials and processes for atmospheric water harvesting He has published more than 145 scientific papers indexed in SCOPUS (google scholar h index: 27). He acted as a referee for numerous international journals. He has contributed to the organization of numerous national and international conferences and was a member of scientific committees for several international conferences. He is supervising PhD, MSc as well as BSc students in the field of PV and CSP. He is an expert of the CNRST in the field of renewable energies

J. Nandha Gopal working as Associate Professor in the Department of Electrical and Electronics Engineering at Velammal Institute of Technology, Tamilnadu, India. He has been teaching in the Electrical and Electronics Engineering since 2006. He received his PhD from College of Engineering, Guindy, Anna University Chennai. His research interest includes Power Electronics, Design of Power Converters, Renewable Energy Systems and Smart Grid. He has published 7 international journals, 8 national and international conference papers, 12 Book chapters and 1 patent. He is a member of Institution of Engineers (I). He is acting as reviewer for various journals like Circuit World, International Journal of Electronics, Measurement and Control, International Journal of Control, Intelligent Automation and Soft Computing, Computers Materials & Continua, Transaction of the Institute of Measurement and Control, Computer Systems Science and Engineering. He has acted as resource person for various institutions.. He as obtained National Instruments customer Education Certification and now granted the title of Certified Lab VIEW Associate Developer (CLAD). He has Received Research Supervisor Recognition under Anna University, Chennai.

Antonio Lecuona-Neumann is an Aeronautical Engineer in 1975. Ph.D. in 1980 on Aeronautical Engineering on laser-driven Nuclear Fusion Associate Professor in 1975. Full Professor in 1993 at the Universidad Carlos III de Madrid, Spain (UC3M).

Honored by the highest Education award in Spain, “Encomienda de Alfonso X El Sabio” in 1996. Been commissioned as Vice-Chancellor and Head of Department. Author of around 100 publications, most of them in high-impact journals. H25. More than 1500 citations. Participated in 40 competitive research projects, including 10 with the EU Commission. Author of 13 patents and more than 100 congress talks. Tutor of 13 Ph.D. theses and more than 100 Master's theses. Member of the Editorial Board of several journals, including “Applied Energy”. Entitled and visiting Professor of master’s degrees in several Universities, Spain and Cabo Verde. Visiting Professor at Stanford University, Dept. Mechanical Engineering HGTL laboratories. 1986/87 Director of the research Institute “Pedro Juan de Lastanosa” at UC3M up to 2023. He acted as an advisor for the Ministry of Science and Education of Spain. Member of the European Advisory Board in July 2005 and of the scientific board in 2006 of the “Centro de Tecnologías Avanzadas para la Combustión Limpia del Carbón” CIEMAT, Spain. Member of the EU Academy of Sciences in 2018. Emeritus Professor in 2023

Rachid Malek is a Full Professor at Mohammed Premier University (Oujda/ Morocco) - National School of Applied Sciences. Prof. Dr. MALEK does research in Renewable Energies and Materials Science. His current projects are Photovoltaic Cells & Systems and Nanotechnologies.

I**Rachid Safoui** studied materials and renewable energy at Ibn Zohr University, in Agadir, Morocco, where he obtained his BS degree in 2018 and MS degree in 2021. He is persuing his PhD degree at the same University under the supervision of Prof. Ahmed Ihlal. His current research focuses on materials design and engineering for atmospheric water harvesting and seawater desalination.

Index

T

www.ingramcontent.com/pod-product-compliance
Lightning Source LLC
LaVergne TN
LVHW080857300926
849162LV00003B/450
* 9 7 9 8 3 3 7 3 2 5 3 5 4 *